Variants of SBC Process Algebra

-- The Structure-Behavior Coalescence Approach --

William S. Chao

3

Structure-Behavior Coalescence

Systems Architecture **=** **Systems Structure** **+** **Systems Behavior**

4

CONTENTS

6

8

ABOUT THE AUTHOR

Dr. William S. Chao is the CEO & founder of SBC Architecture International®. SBC (Structure-Behavior Coalescence) architecture is a systems architecture which demands the integration of systems structure and systems behavior of a system. SBC architecture applies to hardware architecture, software architecture, enterprise architecture, knowledge architecture, and thinking architecture. The core theme of SBC architecture is: "Architecture = Structure + Behavior."

William S. Chao received his bachelor degree (1976) in telecommunication engineering and master degree (1981) in information engineering, both from the National Chiao-Tung University, Taiwan. From 1976 till 1983, he worked as an engineer at Chung-Hwa Telecommunication Company, Taiwan.

William S. Chao received his master degree (1985) in information science and Ph.D. degree (1988) in information science, both from the University of Alabama at Birmingham, USA. From 1988 till 1991, he worked as a computer scientist at GE Research and Development Center, Schenectady, New York, USA.

Dr. William S. Chao has been teaching at National Sun Yat-

Sen University, Taiwan since 1992 and now serves as the president of Association of Enterprise Architects, Taiwan Chapter. His research covers: systems architecture, hardware architecture, software architecture, enterprise architecture, knowledge architecture, and thinking architecture.

PART I: WHAT IS PROCESS ALGEBRA?

Algebraic Approach to the Study of Concurrent Systems

Process algebras are a diverse family of related approaches to the study of concurrent systems. Their tools are algebraic languages for the high-level description of interactions, communications, and synchronizations between a collection of independent agents or processes.

Process algebras also provide algebraic laws that allow process descriptions to be manipulated and analyzed, and permit formal reasoning about equivalences and observation congruence among processes.

Examples of Process Algebras

There are several leading algebraic approaches to modeling concurrent systems.

Communicating Sequential Processes (CSP) was first described in a 1978 paper by C. A. R. Hoare.

Arthur John Robin Gorell Milner introduced the Calculus of Communicating Systems (CCS) around 1980.

Algebra of Communicating Processes (ACP) was initially developed by Jan Bergstra and Jan Willem Klop in 1982.

SBC Process Algebra

SBC process algebra evolved from CCS (Calculus of Communicating Systems).

CCS is a general process algebra language for the study of concurrent systems. Unlike CCS, SBC process algebra is only applicable to systems architecture.

Single-queue SBC process algebra, multi-queue SBC process algebra, and infinite-queue SBC process algebra are the three SBC process algebras.

16

PART II: MATHEMATICS OF SBC PROCESS ALGEBRA

18

Interaction

An interaction represents an indivisible and instantaneous handshake or communication between two agents. The caller agent (either external environment's actor or component) communicates with the callee agent (component) through the interaction.

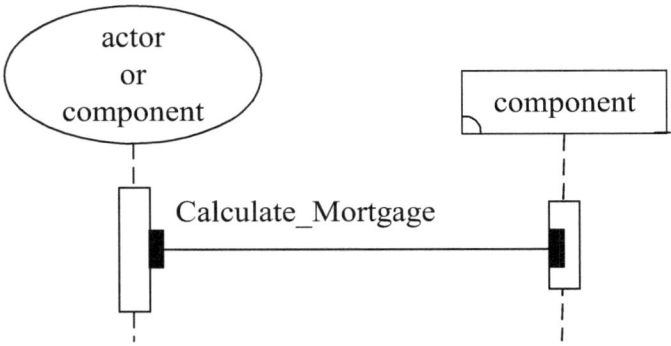

There are two ports, i.e., calling port or called port, of an interaction. The caller agent owns the "calling port" of the interaction.

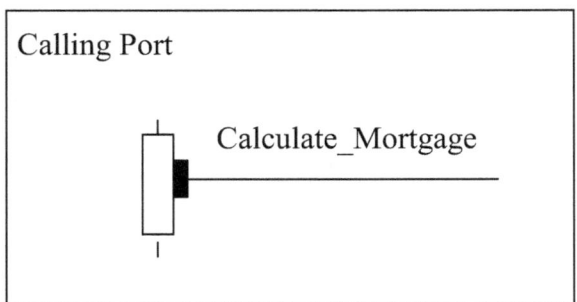

The caller agent together with the "calling port" is named the "calling action".

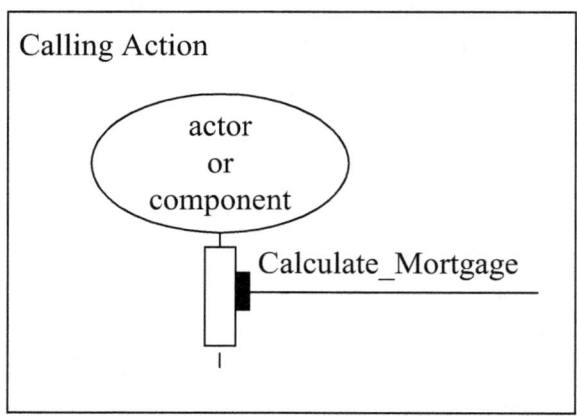

The callee agent owns the "called port" of the interaction.

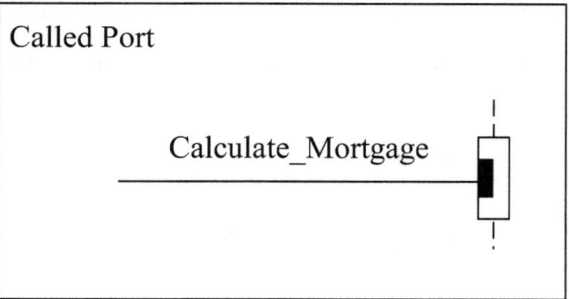

The callee agent together with the "called port" is named the "called action".

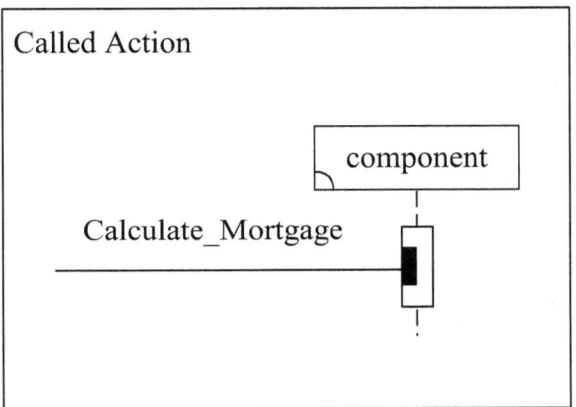

In order to simplify the interaction diagram, we will redraw it as follows.

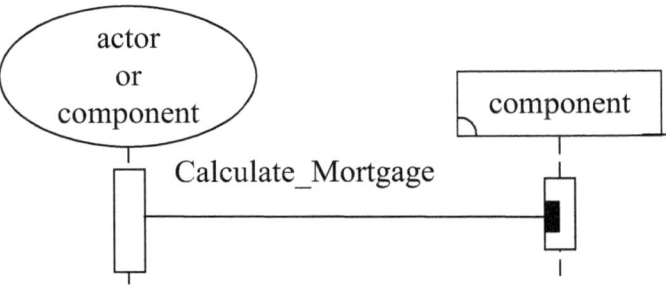

Sequentialization of Interactions

Sometimes interactions must be temporally ordered. For example, it might be desirable to specify algorithms such as: execute the "a" interaction first and then execute the "P" process later. Sequentialization of interactions can be used for such purposes.

Sequentialization of interactions, usually written as the a$\bullet P$ process, indicates that it will perform the "a" interaction first and continue as the "P" process.

Summation of Processes

The binary operator "+", summation, combines two process expressions as alternatives.

For example, the process $P+Q$ can proceed non-deterministically either as the process P or the process Q; as soon as one performs its first interaction the other is discarded.

Parallel Composition of Processes

Parallel composition of two processes P and Q, usually written $P\|Q$, is the key primitive distinguishing the process algebras from sequential models of process executions.

Parallel composition allows the executions in P and Q to proceed simultaneously and independently.

Recursive Definition of a Process

The operators presented so far describe only finite interaction and are consequently insufficient for full computability, which includes non-terminating behavior. Recursion is the operator that allows finite descriptions of infinite behavior.

For example, $\mathbf{fix}(X{=}E)$ can be understood as abbreviating the recursive definition of an infinite behavior denoted by the "X" process variable.

Replication of a Process

Replication is the other operator that allows finite descriptions of infinite behavior of a process.

For example, replication $!P$ can be understood as abbreviating the parallel composition of a countably infinite number of P processes.

Conditional Definition of a Process

A process can be defined by a one-or-more-armed conditional expression.

For example, the process (**if** $cond_1$ **then** P_1)+(**if** $cond_2$ **then** P_2)...+(**if** $cond_j$ **then** P_j) will proceed as the process P_1 if the "$cond_1$" value is true, or proceed as the process P_2 if the "$cond_2$" value is true,..., or proceed as the process P_j if the "$cond_j$" value is true.

Null Process

Process algebras generally also include a null process, denoted as *STOP*, which has no interaction points. It is utterly inactive and its sole purpose is to act as the inductive anchor on top of which more interesting processes can be generated.

The process "*STOP•P₁*" (i.e. sequential composition of processes *STOP* and P_1) equals to the process "*STOP*".

$$STOP \bullet P_1 \ = \ STOP$$

The process "*P₂+STOP*" (i.e. summation of processes P_2 and *STOP*) equals to the process "*STOP+P₂*" (i.e. summation of processes *STOP* and P_2) which equals to the process "*P₂*".

$$P_2 + STOP \ = \ STOP + P_2 \ = \ P_2$$

The process "$P_3 \| STOP$" (i.e. parallel composition of processes P_3 and $STOP$) equals to the process "$STOP \| P_3$" (i.e. parallel composition of processes $STOP$ and P_3) which equals to the process "P_3".

$$P_3 \| STOP \quad = \quad STOP \| P_3 \quad = \quad P_3$$

30

PART III: THE STRUCTURE-BEHAVIOR COALESCENCE APPROACH

Structure-Behavior Coalescence Means to Integrate the Systems Structure and Systems Behavior

Systems structure and systems behavior are the two most prominent views of a system, integrating the systems structure and systems behavior is apparently the best way to achieve an integrated whole of a system.

If we are not able to integrate the systems structure and systems behavior, then there is no way that we are able to integrate the whole system.

Structure-behavior coalescence (SBC) provides an elegant way to integrate the systems structure and systems behavior of a system. In other words, SBC facilitates an integrated whole of a system.

Interactions among Components and Actors to Draw Forth the Systems Behavior

In a system, if the components, and among them and the external environment's actors to interact (or handshake), these interactions will draw forth the systems behavior.

We conclude that "interaction" plays an important factor in integrating the systems structure and systems behavior for a system.

The overall behavior of a system consists of many individual behaviors. Each individual behavior represents an execution path. We use an interaction flow diagram (IFD) to demonstrate this individual behavior.

Collection of All Interaction Flow Diagrams Defines the Systems Architecture

The collection of all interaction flow diagrams defines the integration of systems structure and systems behavior of a system.

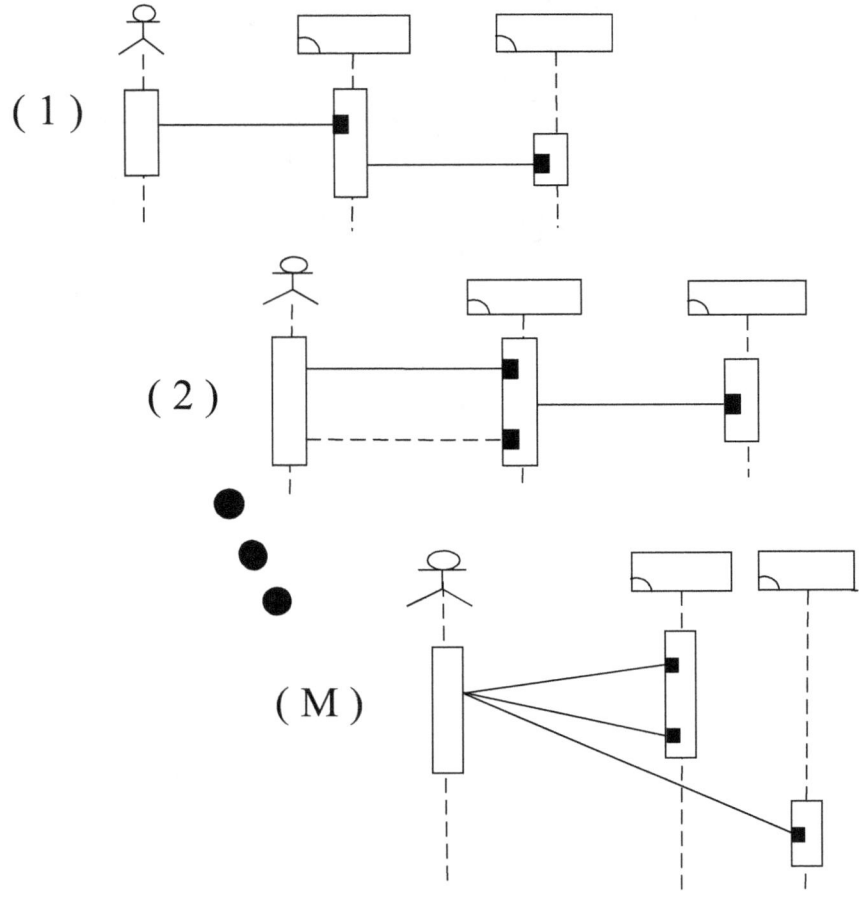

That is, the collection of all interaction flow diagrams defines the systems architecture.

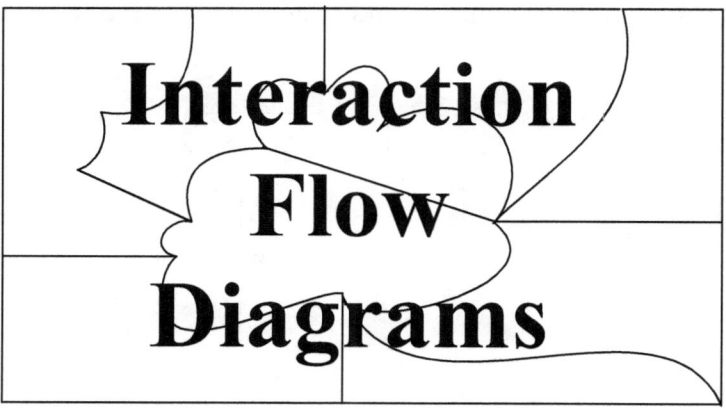

Systems architecture (SA) represents a knowledge repository of a system. Stakeholders can submit and acquire knowledge to and from this knowledge repository.

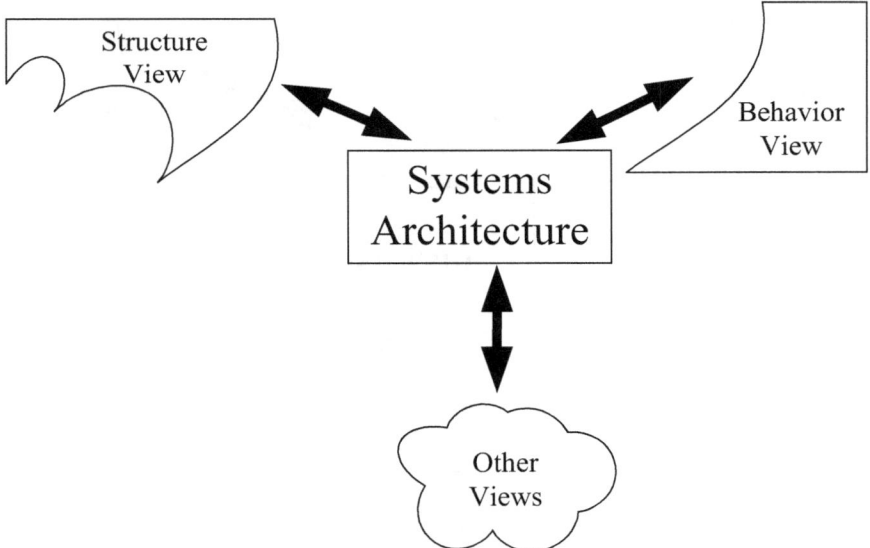

So, the collection of all interaction flow diagrams represents a knowledge repository of a system. Stakeholders can submit and acquire knowledge to and from this knowledge repository.

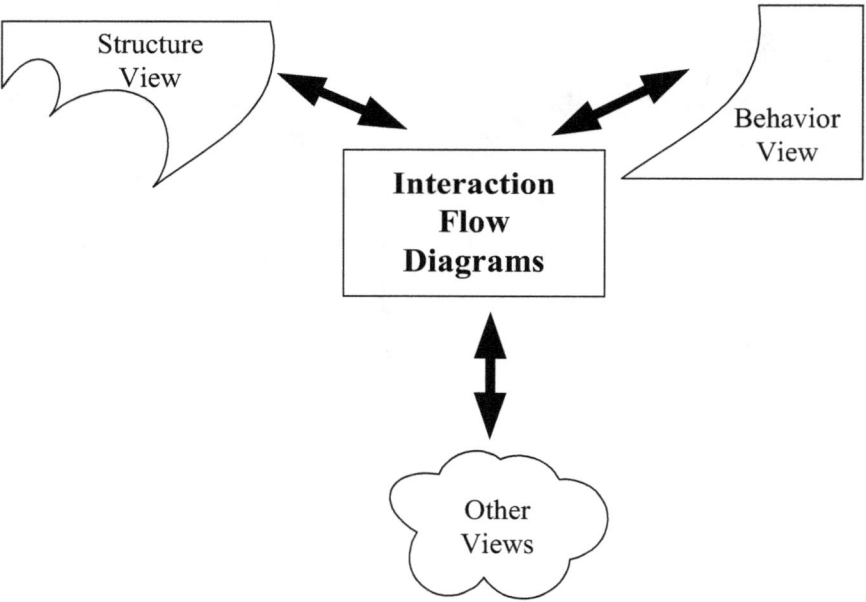

Systems Architecture of the *Sweepstakes Sales Promotion* System

The collection of all interaction flow diagrams defines the systems architecture. The overall behavior of the *Sweepstakes Sales Promotion* system includes two behaviors: *Get_Sweepstake_Number* and *Draw_Out_the_Winners_List*. Each of them is described by an individual IFD.

Customer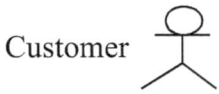

Get_Sweepstake_Number

Draw_Out_the_Winners_List

Sweepstakes Sales Promotion System

An IFD of the *Get_Sweepstake_Number* behavior is shown below. First, actor *Customer* interacts with the *Clerk* component through the *Buy_More_Than_Thousand_Dollars* operation call interaction, carrying the *Sweepstake_Number* output parameter. Last, actor *Customer* interacts with the *Sweepstake_GUI* component through the *Register* operation call interaction, carrying the *Personal_Data* and *Sweepstake_Number* input parameters.

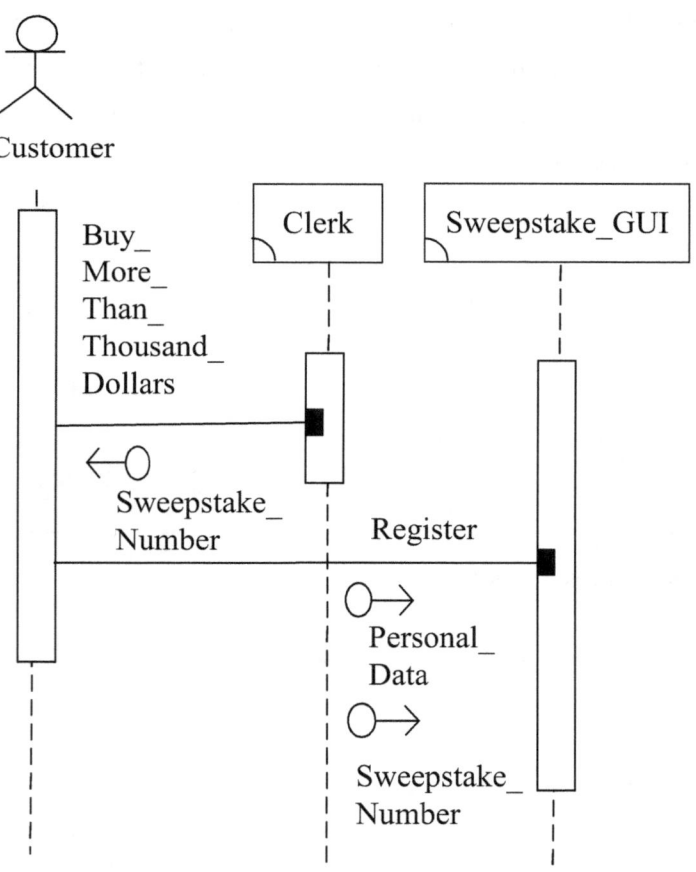

An IFD of the *Draw_Out_the_Winners_List* behavior is shown below. First, actor *Customer* interacts with the *Sweepstake_Ho*st component through the *Draw_Out* operation call interaction. Next, component *Sweepstake_Host* interacts with the *Sweepstake_GUI* component through the *Lucky_Draw* operation call interaction, carrying the *Winners_List* output parameter. Last, actor *Customer* interacts with the *Sweepstake_Host* component through the *Draw_Out* operation return interaction, carrying the *Winners_List* output parameter.

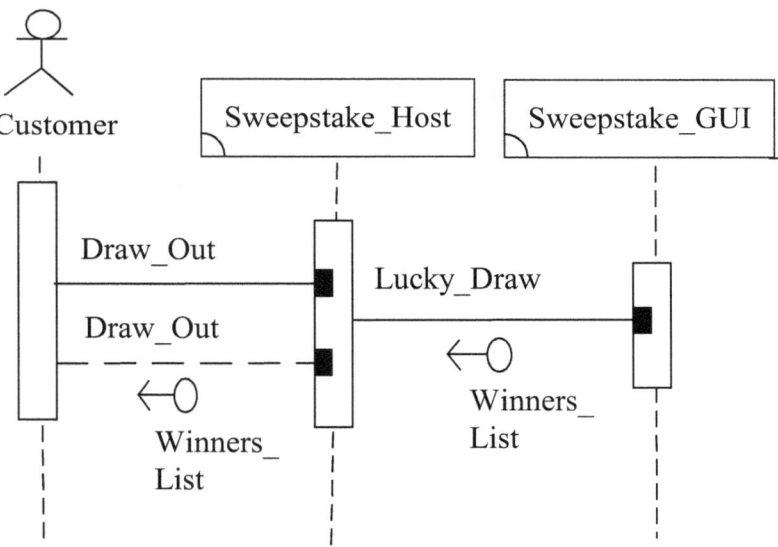

Interaction$_{11}$ stands for the 1st interaction of the 1st interaction flow diagram of the *Sweepstakes Sales Promotion* system. Interaction$_{11}$ describes the *Customer* actor interacts with the *Clerk* component.

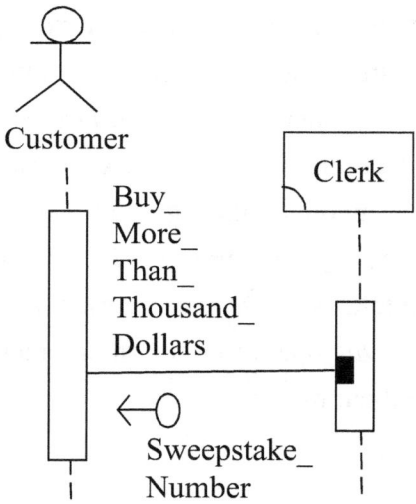

Interaction$_{12}$ stands for the 2nd interaction of the 1st interaction flow diagram of the *Sweepstakes Sales Promotion* system. Interaction$_{12}$ describes the *Customer* actor interacts with the *Sweepstake_GUI* component.

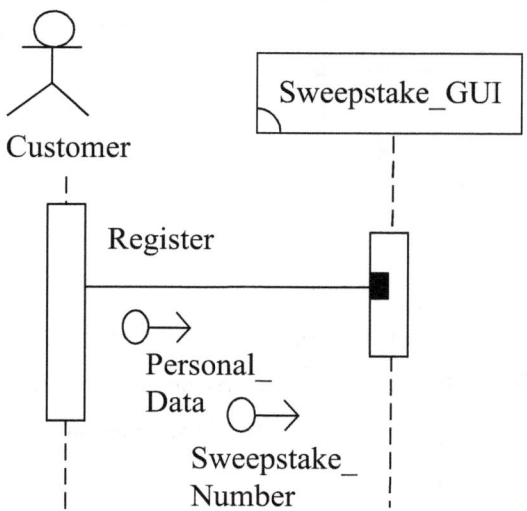

Interaction$_{21}$ stands for the 1st interaction of the 2nd interaction flow diagram of the *Sweepstakes Sales Promotion* system. Interaction$_{21}$ describes the *Customer* actor interacts with the *Sweepstake_Host* component.

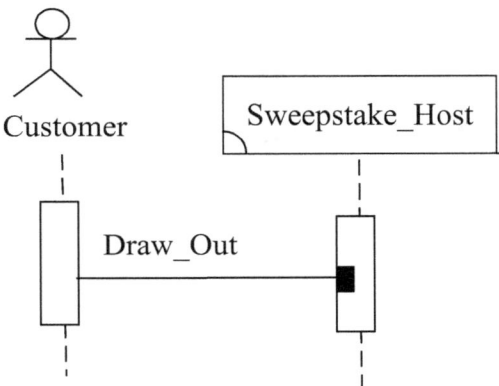

Interaction$_{22}$ stands for the 2nd interaction of the 2nd interaction flow diagram of the *Sweepstakes Sales Promotion* system. Interaction$_{22}$ describes the *Sweepstake_Host* component interacts with the *Sweepstake_GUI* component.

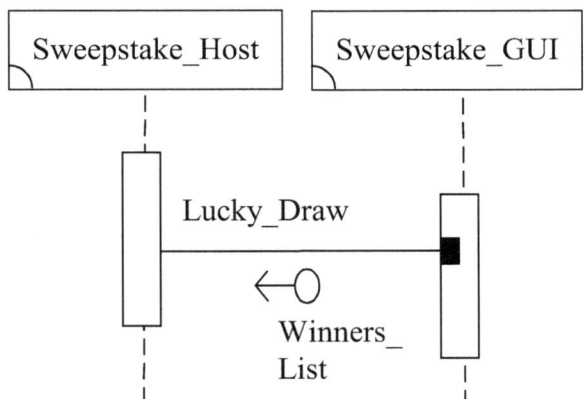

Interaction$_{23}$ stands for the 3rd interaction of the 2nd interaction flow diagram of the *Sweepstakes Sales Promotion* system. Interaction$_{23}$ describes the *Customer* actor interacts with the *Sweepstake_Host* component.

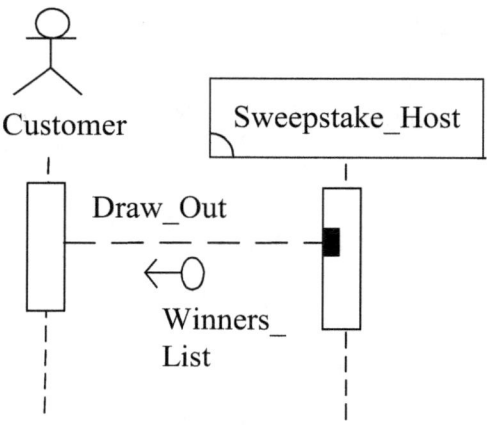

PART IV: EXECUTION OF THE SYSTEMS ARCHITECTURE

Sequential Execution of Interactions of an Interaction Flow Diagram

An interaction flow diagram may consist of many interactions. In this interaction flow diagram, each interaction will be sequentially executed.

The yth execution of the xth interaction flow diagram is defined as a sequence INTERACTIONEXECUTION$_{xy}$ = (Interaction$_{xyz}$) $_{z = 1 \text{ to } N}$, where "z" stands for the zth (z = 1 to N) interaction of the xth interaction flow diagram.

As a first example, the yth execution of the 1st interaction flow diagram of the *Sweepstakes Sales Promotion* system is defined as a sequence INTERACTIONEXECUTION$_{1y}$ = (Interaction$_{1y1}$, Interaction$_{1y2}$).

As a second example, the yth execution of the 2nd interaction flow diagram of the *Sweepstakes Sales Promotion* system is defined as a sequence INTERACTIONEXECUTION$_{2y}$ = (Interaction$_{2y1}$, Interaction$_{2y2}$, Interaction$_{2y3}$).

Infinite Executions of an Interaction Flow Diagram

An interaction flow diagram may be executed an infinite number of times (through recursion or replication).

Infinite executions of the xth interaction flow diagram is defined as a sequence $\text{INTERACTIONEXECUTION}_x$ = $(\text{Interaction}_{xyz})$ $_{y = 1 \text{ to } \infty \text{ and } z = 1 \text{ to } N}$, where "y" stands for the yth (y = 1 to ∞) execution of the xth interaction flow diagram; "z" stands for the zth (z = 1 to N) interaction of the xth interaction flow diagram.

As a first example, infinite executions of the 1st interaction flow diagram of the *Sweepstakes Sales Promotion* system is defined as a sequence $\text{INTERACTIONEXECUTION}_1$ which may look like $(\text{Interaction}_{111}, \text{Interaction}_{112}, \text{Interaction}_{121}, \text{Interaction}_{122}, \text{Interaction}_{131}, \text{Interaction}_{132}, \text{Interaction}_{141}, \text{Interaction}_{142}, \text{Interaction}_{151}, \text{Interaction}_{152}, \ldots, \text{Interaction}_{1\infty1}, \text{Interaction}_{1\infty2})$ if the recursion is used, or look like $(\text{Interaction}_{111}, \text{Interaction}_{121}, \text{Interaction}_{122}, \text{Interaction}_{112}, \text{Interaction}_{131}, \text{Interaction}_{132}, \text{Interaction}_{141}, \text{Interaction}_{142}, \text{Interaction}_{151}, \text{Interaction}_{152}, \ldots, \text{Interaction}_{1\infty1}, \text{Interaction}_{1\infty2})$ if the replication is used.

As a second example, infinite executions of the 2nd interaction flow diagram of the *Sweepstakes Sales Promotion* system is defined as a sequence INTERACTIONEXECUTION$_2$ which may look like (Interaction$_{211}$, Interaction$_{212}$, Interaction$_{213}$, Interaction$_{221}$, Interaction$_{222}$, Interaction$_{223}$, Interaction$_{231}$, Interaction$_{232}$, Interaction$_{233}$, Interaction$_{241}$, Interaction$_{242}$, Interaction$_{243}$, Interaction$_{251}$, Interaction$_{252}$, Interaction$_{253}$,..., Interaction$_{2\infty1}$, Interaction$_{2\infty2}$, Interaction$_{2\infty3}$) if the recursion is used, or look like (Interaction$_{211}$, Interaction$_{212}$, Interaction$_{221}$, Interaction$_{222}$, Interaction$_{223}$, Interaction$_{213}$, Interaction$_{231}$, Interaction$_{232}$, Interaction$_{233}$, Interaction$_{241}$, Interaction$_{242}$, Interaction$_{243}$, Interaction$_{251}$, Interaction$_{252}$, Interaction$_{253}$,..., Interaction$_{2\infty1}$, Interaction$_{2\infty2}$, Interaction$_{2\infty3}$) if the replication is used.

All Infinitely-Executed Individual Interaction Flow Diagrams Tend to Be Non-Deterministically or Parallelly Executed

The systems architecture is a collection of all its individual interaction flow diagrams.

These infinitely-executed individual interaction flow diagrams are mutually independent of each other. They tend to be non-deterministically or parallelly executed.

The non-deterministic or parallel execution of all infinitely-executed individual interaction flow diagrams means to define the execution of the systems architecture.

Execution of the Systems Architecture

The systems architecture may consist of many interaction flow diagrams. Each interaction flow diagram may be executed an infinite number of times. An interaction flow diagram may consist of many interactions.

The execution of the systems architecture is defined as a sequence INTERACTIONEXECUTION = (interaction$_{xyz}$) $_{x = 1 \text{ to } M}$ and $_{y = 1 \text{ to } \infty}$ and $_{z = 1 \text{ to } N}$, where "x" stands for the xth (x = 1 to M) interaction flow diagram of this systems architecture; "y" stands for the yth (y = 1 to ∞) execution of the xth interaction flow diagram; "z" stands for the zth (z = 1 to N) interaction of the xth interaction flow diagram.

For example, there are three possible ways to execute the *Sweepstakes Sales Promotion* system.

INTERACTIONEXECUTION^{SQM} = (interaction$_{111}$, interaction$_{112}$, interaction$_{121}$, interaction$_{122}$, interaction$_{131}$, interaction$_{132}$, interaction$_{211}$, interaction$_{212}$, interaction$_{213}$, interaction$_{141}$, interaction$_{142}$, interaction$_{221}$, interaction$_{222}$, interaction$_{223}$,....) is the first possible execution of this systems architecture.

INTERACTIONEXECUTION^{MQM} = (interaction$_{111}$, interaction$_{211}$, interaction$_{212}$, interaction$_{213}$, interaction$_{112}$, interaction$_{121}$, interaction$_{122}$, interaction$_{131}$, interaction$_{132}$, interaction$_{141}$, interaction$_{142}$, interaction$_{221}$, interaction$_{222}$, interaction$_{223}$,...) is the second possible execution of this systems architecture.

INTERACTIONEXECUTION^{IQM} = (interaction$_{111}$, interaction$_{121}$, interaction$_{131}$, interaction$_{112}$, interaction$_{122}$, interaction$_{132}$, interaction$_{211}$, interaction$_{212}$, interaction$_{213}$, interaction$_{141}$, interaction$_{142}$, interaction$_{221}$, interaction$_{222}$, interaction$_{223}$,...) is third possible execution of this systems architecture.

Interactions Scheduling Defines the Execution of the Systems Architecture

The execution of the systems architecture is described by the scheduling of all interactions.

Different scheduling of all interactions will generate different sequences for executing the systems architecture.

There are at least three models (i.e. SQM, MQM, IQM) of interaction scheduling for the execution of the systems architecture. We will elaborate on their details in the following chapters.

Fairness of Interactions Scheduling

No matter what model we will use for scheduling of interactions, fairness is an important feature that any interactions scheduling algorithm must possess.

Fairness guarantees that every executable interaction will and shall eventually be executed.

SBC Process Algebra

SBC process algebra evolved from CCS (Calculus of Communicating Systems).

CCS is a general process algebra language for the study of concurrent systems. Unlike CCS, SBC process algebra is only applicable to systems architecture.

SBC Process Algebra Defines the Interactions Scheduling

Interactions scheduling defines the execution of the systems architecture. Single-queue model, multi-queue model, and infinite-queue model are the three interaction scheduling models.

SBC process algebra defines the interactions scheduling. Single-queue SBC process algebra, multi-queue SBC process algebra, and infinite-queue SBC process algebra are the three SBC process algebras.

Single-queue SBC process algebra defines the single-queue interactions scheduling. Multi-queue SBC process algebra defines the multi-queue interactions scheduling. Infinite-queue SBC process algebra defines the infinite-queue interactions scheduling.

PART V: SINGLE-QUEUE MODEL
FOR INTERACTIONS
SCHEDULING

Approach of Single-Queue Model

Single-queue model (SQM) adopts a single queue interactions scheduling algorithm. There exists only one queue in this single-queue model.

Whenever the yth execution of the xth interaction flow diagram, IFD_x, is ready, its interactions such as $interaction_{xy1}$, $interaction_{xy2}$,..., $interaction_{xyN}$ will be one-by-one added at the end of the INTERACTION_QUEUE queue. The *Ready_Head* is a pointer variable which points to the interaction at the head of the queue.

60

Ready_Head

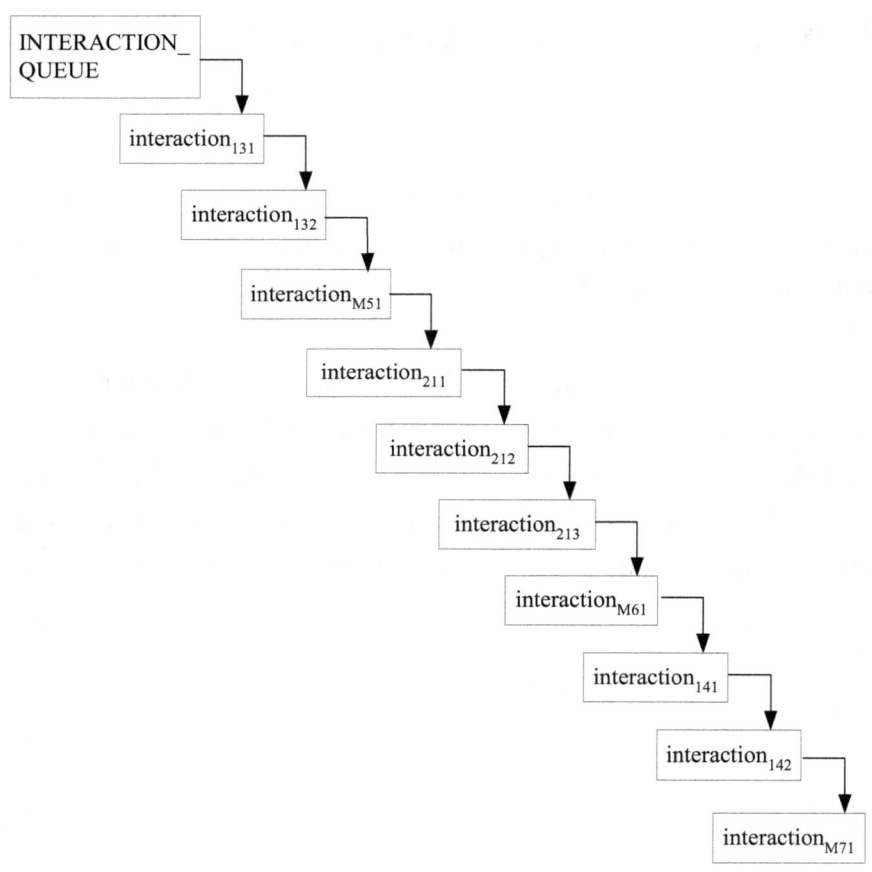

Given the queue structure just described, the SQM scheduling algorithm is simple. It just picks the interaction at the head of that queue. If the queue is empty, then the idle routine will be executed.

Features of Single-Queue Model

The sequence generated by the single-queue model for interactions scheduling algorithm possesses the following characteristics:

(1) It is a fair scheduling. Every executable interaction will and shall eventually be executed.

(2) If $e < f$, then the $interaction_{xye}$ must be executed before the $interaction_{xyf}$.

(3) If $a < b$, then the $interaction_{xae}$ must be executed before the $interaction_{xbf}$.

(4) If $e < f$, then only those $interaction_{xyg}$ which satisfy the "$e < g < f$" condition are allowed to be executed before the $interaction_{xyf}$ and after the $interaction_{xye}$.

PART VI: SINGLE-QUEUE SBC

PROCESS ALGEBRA

Language Constructs of Single-Queue SBC Process Algebra

The set of single-queue SBC process for systems architecture is defined by the following BNF grammar:

(1) <Single-Queue_SBC_Process_of_Systems_Architecture> ::=
   "**fix**(" <Process_Variable> "="<Summation_of_IFD>")"

(2) <Summation_of_IFD> ::=   <IFD>
   | <IFD>"+" <Summation_of_IFD>

(3) <IFD> ::=
     <Type_1_Expression> <Zero_Or_More_Expressions>

(4) <Zero_Or_More_Expressions> ::= "●" <Process_Variable>
   | "●" <Type_1_Or_2_Expression> <Zero_Or_More_Expressions>

(5) <Type_1_Or_2_Expression>   ::= <Type_1_Expression>
   | <Type_2_Expression>

(6) <Type_1_Expression>   ::= <Type_1_Interaction>
   | <Condition> <Type_1_Interaction>
     {"+" <Condition> <Type_1_Interaction>}

(7) <Type_2_Expression>   ::= <Type_2_Interaction>
   | <Condition> <Type_2_Interaction>
     {"+" <Condition> <Type_2_Interaction>}

(8) <Type_1_Interaction> ::= <Actor> <Operation_Call_Or_Return>
             <Operation_Call_Or_Return_Formula> <Component>

(9) <Type_2_Interaction> ::= <Component> <Operation_Call_Or_Return>
             <Operation_Call_Or_Return_Formula> <Component>

Rule 1 describes that the recursion (i.e. **fix**) of summation of all interaction flow diagrams (i.e. Summation_of_IFD) defines the single-queue SBC process for systems architecture.

Rule 1
<Single-Queue_SBC_Process_for_Systems_Architecture> ::= "**fix**(" <Process_Variable> "=" <Summation_of_IFD> ")"

Rule 2 describes that we use either a) an interaction flow diagram (i.e. IFD), or b) an interaction flow diagram (i.e. IFD), followed by a summation (i.e. +), and followed by the summation of all interaction flow diagram, to define the summation of all interaction flow diagrams.

Rule 2
<Summation_of_IFD>   ::=   <IFD> | <IFD> "+" <Summation_of_IFD>

Rule 3 describes that an interaction flow diagram (i.e. IFD) consists of a type_1 expression (i.e. Type_1_Expression) and followed by zero or more expressions (i.e. Zero_Or_More_Expressions).

Rule 3
<IFD> ::=   <Type_1_Expression> <Zero_Or_More_Expressions>

Rule 4 describes that zero or more expressions (i.e. Zero_Or_More_Expressions) consist of either a) a sequential composition (i.e. ●) and followed by a process variable, or b) a sequential composition (i.e. ●), followed by a type_1_or_2 expression (i.e. Type_1_Or_2_Expression), and followed by zero or more expressions (i.e. Zero_Or_More_Expressions).

Rule 4
<Zero_Or_More_Expressions> ::=   "● "<Process_Variable> | "● " <Type_1_Or_2_Expression> <Zero_Or_More_Expressions>

Rule 5 describes that the type_1_or_2 expression (i.e. Type_1_Or_2_Expression) is either a type_1 expression (i.e. Type_1_Expression) or a type_2 expression (i.e. Type_2_Expression).

Rule 5
<Type_1_Or_2_Expression> ::=   <Type_1_Expression> | <Type_2_Expression>

Rule 6 describes that the type_1 expression (i.e. Type_1_Expression) is either a type_1 interaction (i.e. Type_1_Interaction) or a conditional type_1 interaction (i.e. one-or-more-armed conditional expression of Type_1_Interaction).

Rule 6

<Type_1_Expression>   ::= <Type_1_Interaction>
  | <Condition> <Type_1_Interaction>
     {"+" <Condition> <Type_1_Interaction>}

Rule 7 describes that the type_2 expression (i.e. Type_2_Expression) is either a type_2 interaction (i.e. Type_2_Interaction) or a conditional type_2 interaction (i.e. one-or-more-armed conditional expression of Type_2_Interaction).

Rule 7
<Type_2_Expression>   ::= <Type_2_Interaction> | <Condition> <Type_2_Interaction>   {"+" <Condition> <Type_2_Interaction>}

Rule 8 describes that an actor interacting with a component defines the type_1 interaction.

Rule 8
<Type_1_Interaction> ::= <Actor> <Operation_Call_Or_Return>   <Operation_Call_Or_Return_Formula> <Component>

Rule 9 describes that a component interacting with another component defines the type_2 interaction.

Rule 9
<Type_2_Interaction> ::= <Component> <Operation_Call_Or_Return> <Operation_Call_Or_Return_Formula> <Component>

Transitional Semantics of Single-Queue SBC Process Algebra

We assume an infinite set Δ of type_1_or_2 interactions, and use a, b to range over Δ. Further, we let X be the set of process variables, and use X, Y to range over X. We let Φ be the set of process Constants, and use A, B to range over Φ. We let Π be the set of processes, and use P, Q to range over Π. We let Ψ be the set of process expressions, and use E, F to range over Ψ.

Entity set	Entity name	Type of entity
Δ	a, b,...	type_1_or_2 interactions
X	X, Y,...	process variables
Φ	A, B,...	process Constants
	I, J,...	indexing sets
Π	P, Q,...	processes
Ψ	E, F,...	process expressions

In giving meaning to the single-queue SBC process algebra, we shall use the following labelled transition system (LTS)

$$(\Psi, \Delta, \rightarrow)$$

which consists of a set Ψ of process expressions, a set Δ of "type 1 or 2 interactions", and a transition relation $\rightarrow \subseteq \Psi \times \Delta \times \Psi$ where $(E_i, a, E_j) \in \rightarrow$ is denoted by $E_i \xrightarrow{a} E_j$.

The semantics for Ψ consists in the transition rules of each transition relation $\rightarrow$ over $\Psi \times \Delta \times \Psi$. These transition rules will follow the structure of expressions.

We give the complete set of transition rules; the names Prefix, Sum, Recursion, and Constant indicate that the rules are associated respectively with Prefix, Summation, Recursion and with Constants.

$$\text{Prefix} \qquad \frac{\rule{3cm}{0.4pt}}{a \bullet E \xrightarrow{a} E}$$

$$\text{Sum}_j \qquad \frac{E_j \xrightarrow{a} E'_j}{\sum_{i \in I} E_i \xrightarrow{a} E'_j} \, (j \in I)$$

$$\text{Recursion} \qquad \frac{E\{\mathbf{fix}(X=E)/X\} \xrightarrow{a} E'}{\mathbf{fix}(X=E) \xrightarrow{a} E'}$$

$$\text{Constant} \qquad \frac{P \xrightarrow{a} P'}{A \xrightarrow{a} P'} \, (A \stackrel{\text{def}}{=} P)$$

The rule for Prefix can be read as follows: Under any circumstances, we always infer $a \bullet E \xrightarrow{a} E$. That is, an expression, with an interaction prefixed to it, will use this interaction to accomplish the transition.

$$\frac{\rule{3cm}{0.4pt}}{a \bullet E \xrightarrow{a} E}$$

The rule for Summation can be read as follows: If any one summand E_j of the sum $\sum_{i \in I}$ has an interaction, then the whole sum also has that interaction.

$$\frac{E_j \xrightarrow{a} E'_j}{\sum_{i \in I} E_i \xrightarrow{a} E'_j} (j \in I)$$

76

The rule for Recursion can be read as follows: This says that any interaction which may be inferred for the **fix** expression 'unwound' once (by substituting itself for its bound variable) may be inferred for the **fix** expression itself.

$$\frac{E\{\mathbf{fix}(X{=}E)/X\} \xrightarrow{a} E'}{\mathbf{fix}(X{=}E) \xrightarrow{a} E'}$$

The rule for Constants can be read as follows: the rule of Constants asserts that each Constant has the same transitions as its defining expression.

$$\frac{P \xrightarrow{a} P'}{A \xrightarrow{a} P'} \quad (A \overset{\mathrm{def}}{=} P)$$

PART VII: SINGLE-QUEUE INTERACTIONS SCHEDULING VS. SINGLE-QUEUE SBC PROCESS ALGEBRA

Single-Queue Interactions Scheduling for the *Sweepstakes Sales Promotion* System

Sweepstakes Sales Promotion system includes two interaction flow diagrams: IFD_1 and IFD_2.

Whenever the yth execution of the xth (x = 1 to 2) interaction flow diagram, $IFD_{x\ (x = 1\ to\ 2)}$, of the *Sweepstakes Sales Promotion* system is ready, its interactions such as $interaction_{xy1}$, $interaction_{xy2}, \ldots$, $interaction_{xyN}$ will be one-by-one added at the end of the INTERACTION_QUEUE queue. The *Ready_Head* is a pointer variable which points to the interaction at the head of the queue.

Ready_Head

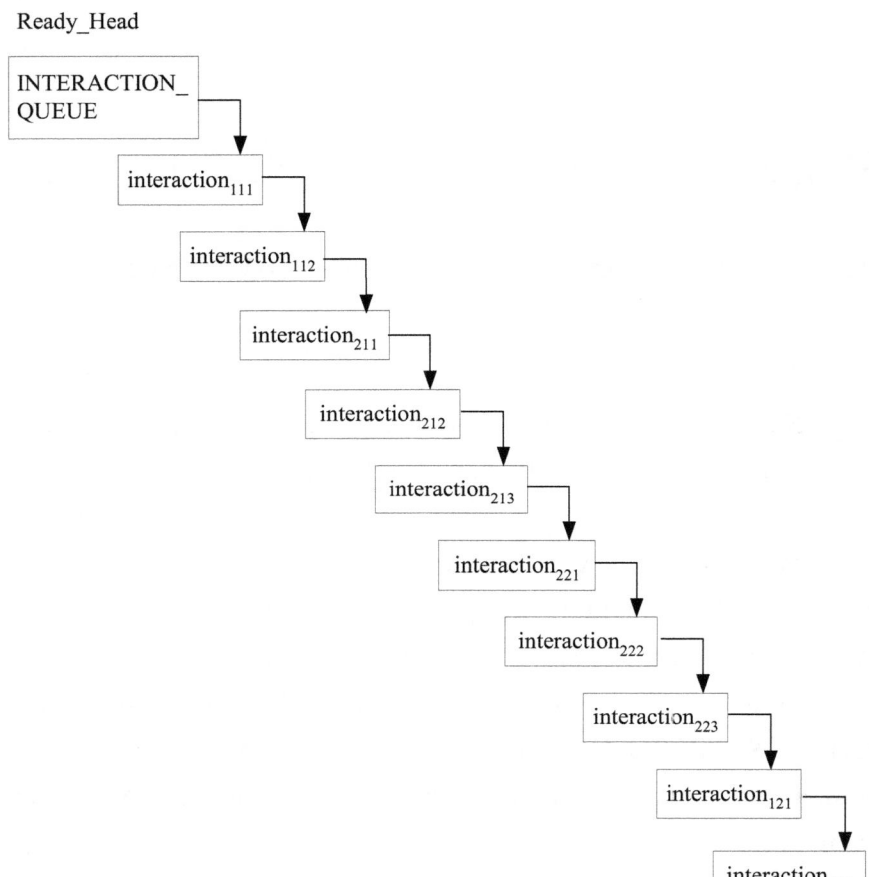

Single-Queue SBC Process of the *Sweepstakes Sales Promotion* System

We draw the single-queue SBC process algebra Backus-Naur Form tree of the *Sweepstakes Sales Promotion* system as follows:

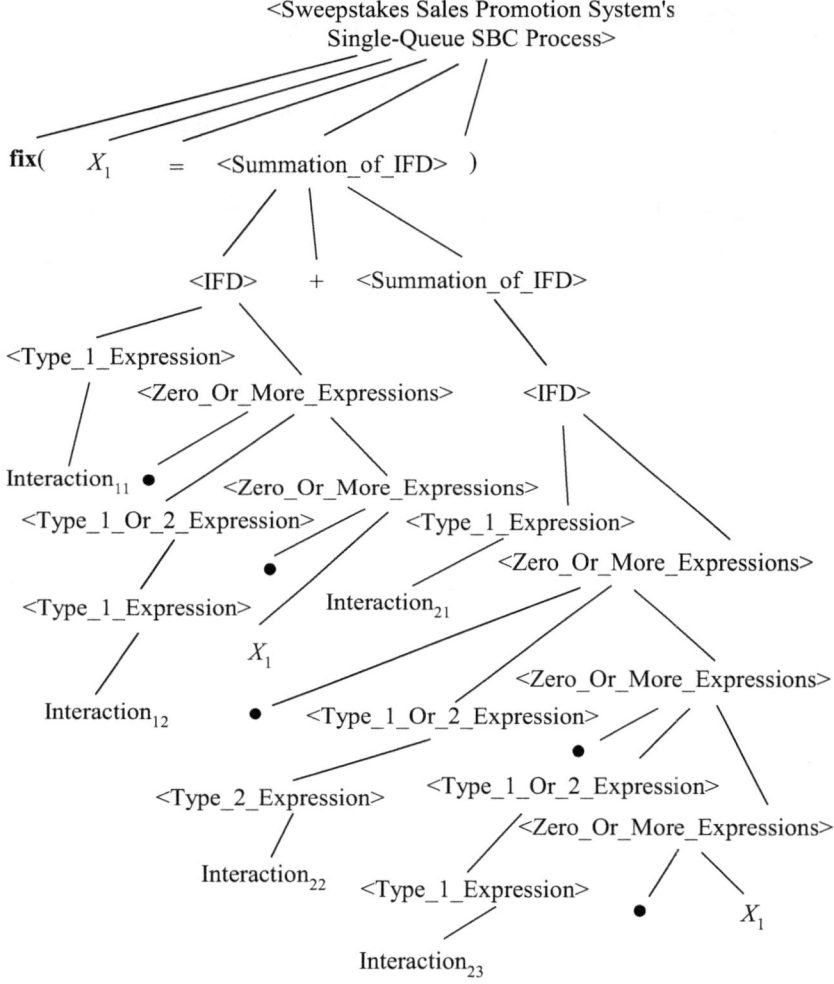

The *Sweepstakes Sales Promotion* system's single-queue SBC process is syntactically represented as "$\mathbf{fix}(X_1=\text{Interaction}_{11}\bullet\text{Interaction}_{12}\bullet X_1+\text{Interaction}_{21}\bullet\text{Interaction}_{22}\bullet\text{Interaction}_{23}\bullet X_1)$".

Sweepstakes Sales Promotion system's Single-Queue SBC Process $\overset{\text{def}}{=\!=}$

$\mathbf{fix}(X_1 = \text{Interaction}_{11}\bullet\text{Interaction}_{12}\bullet X_1 +$
$\qquad\qquad \text{Interaction}_{21}\bullet\text{Interaction}_{22}\bullet\text{Interaction}_{23}\bullet X_1)$

We define process P_{01} as the *Sweepstakes Sales Promotion* system's single-queue SBC process. Thereafter, semantics of the *Sweepstakes Sales Promotion* system's single-queue SBC process is demonstrated by the following transition graph.

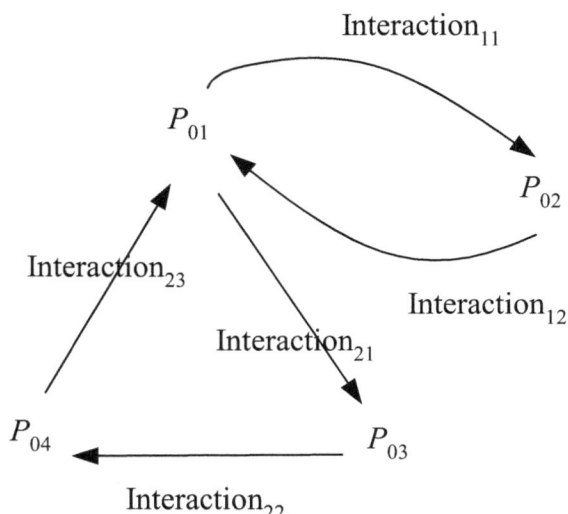

In the transition graph of the *Sweepstakes Sales Promotion* system's single-queue SBC process, processes P_{01}, P_{02}, P_{03}, and P_{04} are defined as:

$$P_{01} \stackrel{\text{def}}{=} \text{Interaction}_{11} \bullet P_{02} + \text{Interaction}_{21} \bullet P_{03}$$

$$P_{02} \stackrel{\text{def}}{=} \text{Interaction}_{12} \bullet P_{01}$$

$$P_{03} \stackrel{\text{def}}{=} \text{Interaction}_{22} \bullet P_{04}$$

$$P_{04} \stackrel{\text{def}}{=} \text{Interaction}_{23} \bullet P_{01}$$

Single-Queue SBC Process Algebra Defines the Single-Queue Interactions Scheduling

Examining both the single-queue model for interactions scheduling and single-queue SBC process algebra, we find out that both of them generate the same interactions execution sequence.

Therefore, we conclude that single-queue SBC process algebra defines the single-queue interactions scheduling.

PART VIII: MULTI-QUEUE MODEL FOR INTERACTIONS SCHEDULING

Approach of Multi-Queue Model

Multi-queue model (MQM) adopts a multiple queues interactions scheduling algorithm. The number of interaction flow diagrams will be used as the number of queues. All queues are treated equally, without any preference.

Whenever the yth execution of the xth interaction flow diagram, IFD_x, is ready, its interactions such as $interaction_{xy1}$, $interaction_{xy2}$, ..., $interaction_{xyN}$ will be one-by-one added at the end of the IFD_x_QUEUE queue. That is, all executions of all interactions of the 1st interaction flow diagram are queued in the IFD_1_QUEUE queue; all executions of all interactions of the 2nd interaction flow diagram are queued in the IFD_2_QUEUE queue; ...; all executions of all interactions of the Mth interaction flow diagram are queued in the IFD_M_QUEUE queue. The array *Ready_Head* has one entry for each queue, with that entry pointing to the interaction at the head of the queue.

Ready_Head

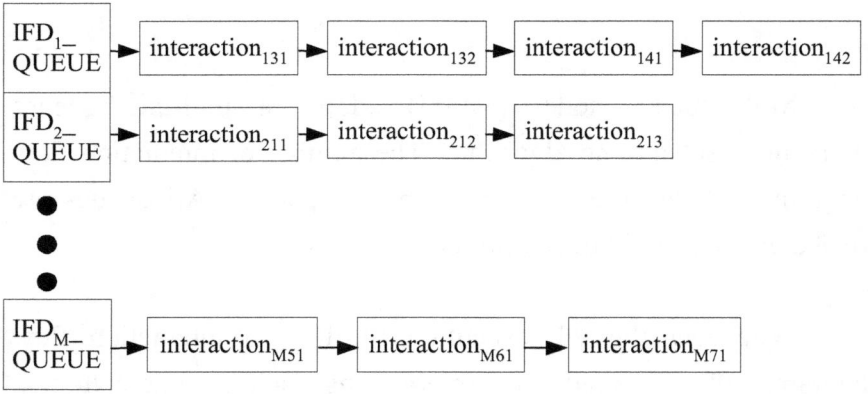

Given the queue structures just described, the MQM scheduling algorithm is simple. It just randomly chooses a queue that is not empty and picks the interaction at the head of that queue. If all the queues are empty, then the idle routine will be executed.

Features of Multi-Queue Model

The sequence generated by the multi-queue model for interactions scheduling algorithm possesses the following characteristics:

(1) It is a fair scheduling. Every executable interaction will and shall eventually be executed.

(2) If $e < f$, then the $interaction_{xye}$ must be executed before the $interaction_{xyf}$.

(3) If $a < b$, then the $interaction_{xae}$ must be executed before the $interaction_{xbf}$.

PART IX: MULTI-QUEUE SBC
PROCESS ALGEBRA

Language Constructs of Multi-Queue SBC Process Algebra

The set of multi-queue SBC process for systems architecture is defined by the following BNF grammar:

(1) <Multi-Queue_SBC_Process_of_Systems_Architecture> ::=
    <Parallel_of_FixIFD>

(2) <Parallel_of_FixIFD> ::= *STOP*
  | **fix**("<Process_Variable>"="<IFD><Process_Variable>")"
    "‖" <Parallel_of_FixIFD>

(3) <IFD> ::=
    <Type_1_Expression> <Zero_Or_More_Expressions>

(4) <Zero_Or_More_Expressions> ::= " • "
  | " • " <Type_1_Or_2_Expression> <Zero_Or_More_Expressions>

(5) <Type_1_Or_2_Expression> ::= <Type_1_Expression>
  | <Type_2_Expression>

(6) <Type_1_Expression> ::= <Type_1_Interaction>
  | <Condition> <Type_1_Interaction>
    {"+" <Condition> <Type_1_Interaction>}

(7) <Type_2_Expression> ::= <Type_2_Interaction>
  | <Condition> <Type_2_Interaction>
    {"+" <Condition> <Type_2_Interaction>}

(8) <Type_1_Interaction> ::= <Actor> <Operation_Call_Or_Return>
    <Operation_Call_Or_Return_Formula> <Component>

(9) <Type_2_Interaction> ::= <Component> <Operation_Call_Or_Return>
    <Operation_Call_Or_Return_Formula> <Component>

Rule 1 describes that the parallel of all recursive interaction flow diagrams (i.e. Parallel_of_FixIFD), in which each interaction flow diagram may loop itself a countably infinite times, defines the multi-queue SBC process for systems architecture.

Rule 1
<Multi-Queue_SBC_Process_for_Systems_Architecture> ::=   <Parallel_of_FixIFD>

Rule 2 describes that we use either a) a null process (i.e. *STOP*), or b) the parallel composition (i.e. ||) composing the recursion (i.e. **fix**) of an interaction flow diagram and the parallel of all recursive interaction flow diagrams (i.e. Parallel_of_FixIFD), to define the parallel of all recursive interaction flow diagrams (i.e. Parallel_of_FixIFD).

Rule 2
<Parallel_of_FixIFD> ::= *STOP* 	**fix**(”<Process_Variable>“=”<IFD><Process_Variable>“)” “||” <Parallel_of_FixIFD>

Rule 3 describes that an interaction flow diagram (i.e. IFD) consists of a type_1 expression (i.e. Type_1_Expression) and followed by zero or more expressions (i.e. Zero_Or_More_Expressions).

Rule 3
<IFD> ::=   <Type_1_Expression> <Zero_Or_More_Expressions>

Rule 4 describes that zero or more expressions (i.e. Zero_Or_More_Expressions) either a) are a sequential composition (i.e. ●), or b) consist of a sequential composition (i.e. ●), followed by a type_1_or_2 expression (i.e. Type_1_Or_2_Expression), and followed by zero or more expressions (i.e. Zero_Or_More_Expressions).

Rule 4
<Zero_Or_More_Expressions> ::=   " ● " | " ● " <Type_1_Or_2_Expression> <Zero_Or_More_Expressions>

Rule 5 describes that the type_1_or_2 expression (i.e. Type_1_Or_2_Expression) is either a type_1 expression (i.e. Type_1_Expression) or a type_2 expression (i.e. Type_2_Expression).

Rule 5
<Type_1_Or_2_Expression> ::=   <Type_1_Expression> | <Type_2_Expression>

Rule 6 describes that the type_1 expression (i.e. Type_1_Expression) is either a type_1 interaction (i.e. Type_1_Interaction) or a conditional type_1 interaction (i.e. one-or-more-armed conditional expression of Type_1_Interaction).

Rule 6

<Type_1_Expression>   ::= <Type_1_Interaction>
  | <Condition> <Type_1_Interaction>
      {"+" <Condition> <Type_1_Interaction>}

Rule 7 describes that the type_2 expression (i.e. Type_2_Expression) is either a type_2 interaction (i.e. Type_2_Interaction) or a conditional type_2 interaction (i.e. one-or-more-armed conditional expression of Type_2_Interaction).

Rule 7
<Type_2_Expression>　　::=　<Type_2_Interaction> | <Condition> <Type_2_Interaction> 　　{"+" <Condition> <Type_2_Interaction>}

Rule 8 describes that an actor interacting with a component defines the type_1 interaction.

Rule 8
<Type_1_Interaction> ::= <Actor> <Operation_Call_Or_Return> 　　　　　<Operation_Call_Or_Return_Formula> <Component>

Rule 9 describes that a component interacting with another component defines the type_2 interaction.

Rule 9
<Type_2_Interaction> ::= <Component> <Operation_Call_Or_Return> <Operation_Call_Or_Return_Formula> <Component>

Transitional Semantics of Multi-Queue SBC Process Algebra

We assume an infinite set Δ of type_1_or_2 interactions, and use a, b to range over Δ. Further, we let X be the set of process variables, and use X, Y to range over X. We let Φ be the set of process Constants, and use A, B to range over Φ. We let Π be the set of processes, and use P, Q to range over Π. We let Ψ be the set of process expressions, and use E, F to range over Ψ.

Entity set	Entity name	Type of entity
Δ	a, b,...	type_1_or_2 interactions
X	X, Y,...	process variables
Φ	A, B,...	process Constants
Π	P, Q,...	processes
Ψ	E, F,...	process expressions

In giving meaning to the multi-queue SBC process algebra, we shall use the following labelled transition system (LTS)

$$(\Psi, \Delta, \rightarrow)$$

which consists of a set Ψ of process expressions, a set Δ of "type 1 or 2 interactions", and a transition relation $\rightarrow \subseteq \Psi X \Delta X \Psi$ where

$(E_i, a, E_j) \in \rightarrow$ is denoted by $E_i \xrightarrow{a} E_j$.

The semantics for Ψ consists in the transition rules of each transition relation $\rightarrow$ over $\Psi X \Delta X \Psi$. These transition rules will follow the structure of expressions.

We give the complete set of transition rules; the names Prefix, Parallel, Recursion, and Constant indicate that the rules are associated respectively with Prefix, Parallel Composition, Recursion and with Constants.

Prefix
$$\frac{}{a \bullet E \overset{a}{\to} E}$$

Parallel$_1$
$$\frac{E \overset{a}{\to} E'}{E \,\|\, F \overset{a}{\to} E' \,\|\, F}$$

Parallel$_2$
$$\frac{F \overset{a}{\to} F'}{E \,\|\, F \overset{a}{\to} E \,\|\, F'}$$

Recursion
$$\frac{E\{\mathbf{fix}(X{=}E)/X\} \overset{a}{\to} E'}{\mathbf{fix}(X{=}E) \overset{a}{\to} E'}$$

Constant
$$\frac{P \overset{a}{\to} P'}{A \overset{a}{\to} P'} \quad (A \overset{\text{def}}{=\!=} P)$$

The rule for Prefix can be read as follows: Under any circumstances, we always infer $a \bullet E \xrightarrow{a} E$. That is, an expression, with an interaction prefixed to it, will use this interaction to accomplish the transition.

$$\frac{}{a \bullet E \xrightarrow{a} E}$$

There are two transition rules for parallel composition. Rule Parallel$_1$ indicates that from $E \xrightarrow{a} E'$ we shall infer $E \| F \xrightarrow{a} E' \| F$.

$$\frac{E \xrightarrow{a} E'}{E \| F \xrightarrow{a} E' \| F}$$

Rule Parallel$_2$ indicates that from $F \xrightarrow{a} F$' we shall infer $E \| F \xrightarrow{a} E \| F$'.

$$\frac{F \xrightarrow{a} F'}{E \| F \xrightarrow{a} E \| F'}$$

The rule for Recursion can be read as follows: This says that any interaction which may be inferred for the **fix** expression 'unwound' once (by substituting itself for its bound variable) may be inferred for the **fix** expression itself.

$$\frac{E\{\mathbf{fix}(X{=}E)/X\} \xrightarrow{a} E'}{\mathbf{fix}(X{=}E) \xrightarrow{a} E'}$$

The rule for Constants can be read as follows: the rule of Constants asserts that each Constant has the same transitions as its defining expression.

$$\frac{P \xrightarrow{a} P'}{A \xrightarrow{a} P'} \quad (A \stackrel{\mathrm{def}}{=\!=} P)$$

PART X: MULTI-QUEUE INTERACTIONS SCHEDULING VS. MULTI-QUEUE SBC PROCESS ALGEBRA

Multi-Queue Interactions Scheduling for the Sweepstakes Sales Promotion System

Sweepstakes Sales Promotion system includes two interaction flow diagrams: IFD_1 and IFD_2.

Whenever the yth execution of the xth (x = 1 to 2) interaction flow diagram, $IFD_{x\ (x = 1\ to\ 2)}$, of the *Sweepstakes Sales Promotion* system is ready, its interactions such as interaction$_{xy1}$, interaction$_{xy2}$,..., interaction$_{xyN}$ will be one-by-one added at the end of the IFD_x_QUEUE queue. That is, all executions of all interactions of the 1st interaction flow diagram are queued in the IFD_1_QUEUE queue; all executions of all interactions of the 2nd interaction flow diagram are queued in the IFD_2_QUEUE queue. The array *Ready_Head* has one entry for each queue, with that entry pointing to the interaction at the head of the queue.

112

Ready_Head

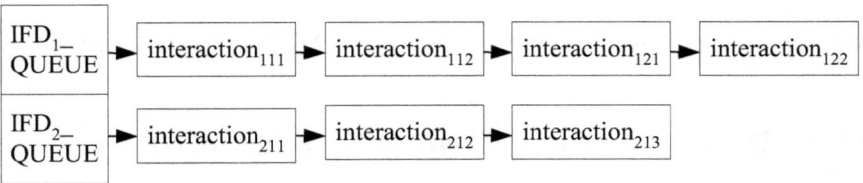

Multi-Queue SBC Process of the Sweepstakes Sales Promotion System

We draw the multi-queue SBC process algebra Backus-Naur Form tree of the *Sweepstakes Sales Promotion* system as follows:

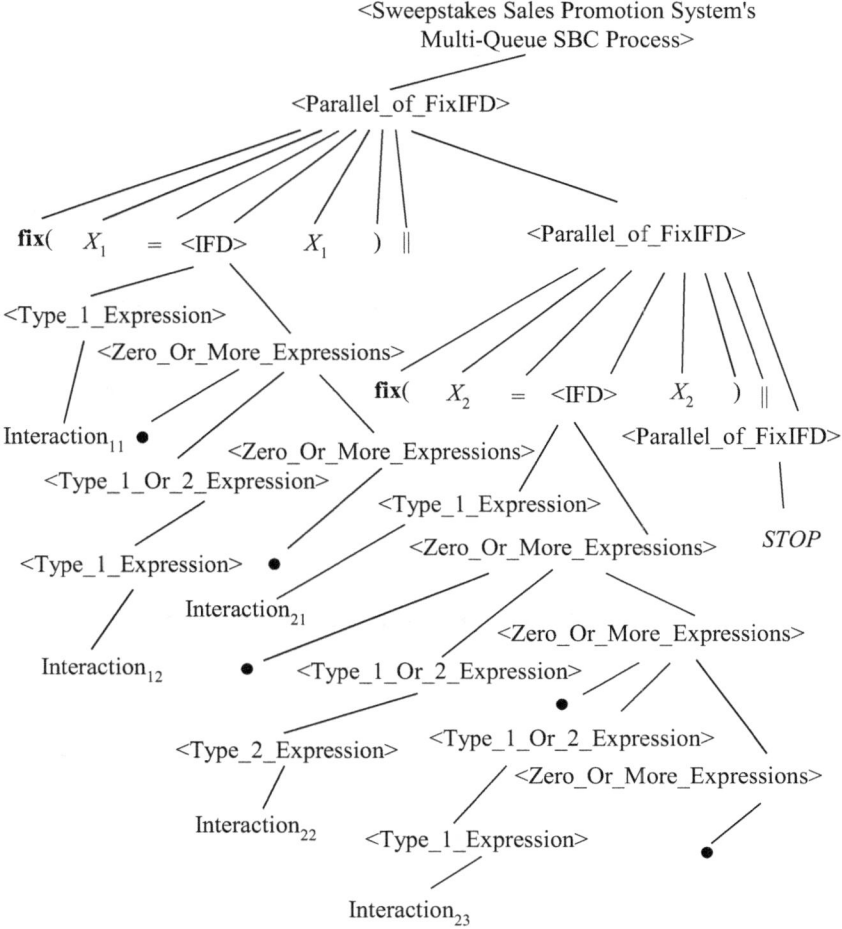

FixIFD$_1$ describes the recursion of the 1st interaction flow diagram, i.e. *Get_Sweepstake_Number* behavior, of the *Sweepstakes Sales Promotion* system. FixIFD$_1$ is syntactically represented as "**fix**(X_1=Interaction$_{11}$•Interaction$_{12}$•X_1)".

$$\text{FixIFD}_1 \overset{\text{def}}{=\!=}$$

$$\textbf{fix}(X_1 = \text{Interaction}_{11} \bullet \text{Interaction}_{12} \bullet X_1)$$

FixIFD$_2$ describes the recursion of the 2nd interaction flow diagram, i.e. *Draw_Out_the_Winners_List* behavior, of the *Sweepstakes Sales Promotion* system. FixIFD$_2$ is syntactically represented   as "**fix**(X_2=Interaction$_{21}$•Interaction$_{22}$•Interaction$_{23}$•X_2)".

$$\text{FixIFD}_2 \overset{\text{def}}{=\!=}$$

$$\textbf{fix}(X_2 = \text{Interaction}_{21} \bullet \text{Interaction}_{22} \bullet \text{Interaction}_{23} \bullet X_2)$$

We define process P_{01} as the FixIFD$_1$ process. Thereafter, semantics of the FixIFD$_1$ process is demonstrated by the following transition graph.

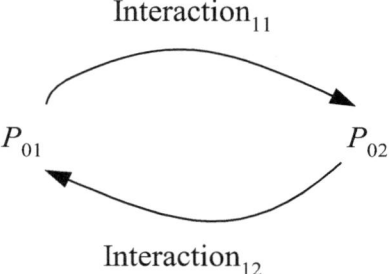

In the transition graph of the FixIFD$_1$ process, processes P_{01} and P_{02} are defined as:

$$P_{01} \overset{\text{def}}{=\!=} \text{Interaction}_{11} \bullet P_{02}$$

$$P_{02} \overset{\text{def}}{=\!=} \text{Interaction}_{12} \bullet P_{01}$$

We define process Q_{01} as the FixIFD$_2$ process. Thereafter, semantics of the FixIFD$_2$ process is demonstrated by the following transition graph.

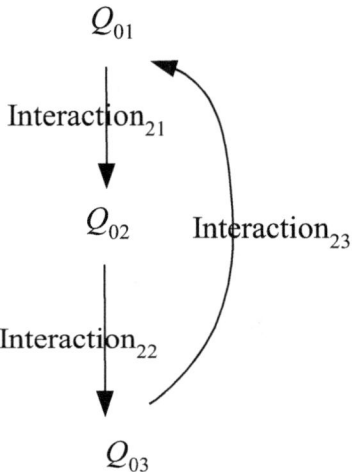

In the transition graph of the FixIFD$_2$ process, processes Q_{01}, Q_{02}, and Q_{03} are defined as:

$$Q_{01} \overset{\text{def}}{=\joinrel=} \text{Interaction}_{21} \bullet Q_{02}$$

$$Q_{02} \overset{\text{def}}{=\joinrel=} \text{Interaction}_{22} \bullet Q_{03}$$

$$Q_{03} \overset{\text{def}}{=\joinrel=} \text{Interaction}_{23} \bullet Q_{01}$$

The *Sweepstakes Sales Promotion* system's multi-queue SBC process is syntactically represented as $\text{FixIFD}_1 \| \text{FixIFD}_2$ which equals to "$\mathbf{fix}(X_1=\text{Interaction}_{11} \bullet \text{Interaction}_{12} \bullet X_1)$ $\|$ $\mathbf{fix}(X_2=\text{Interaction}_{21} \bullet \text{Interaction}_{22} \bullet \text{Interaction}_{23} \bullet X_2)$".

Sweepstakes Sales Promotion system's Multi-Queue SBC Process $\overset{\text{def}}{=\joinrel=}$

$\mathbf{fix}(X_1=\text{Interaction}_{11} \bullet \text{Interaction}_{12} \bullet X_1) \|$
$\mathbf{fix}(X_2=\text{Interaction}_{21} \bullet \text{Interaction}_{22} \bullet \text{Interaction}_{23} \bullet X_2)$

We define process $P_{01} \| Q_{01}$ as the *Sweepstakes Sales Promotion* system's multi-queue SBC process. Thereafter, semantics of the *Sweepstakes Sales Promotion* system's multi-queue SBC process is demonstrated by the following transition graph.

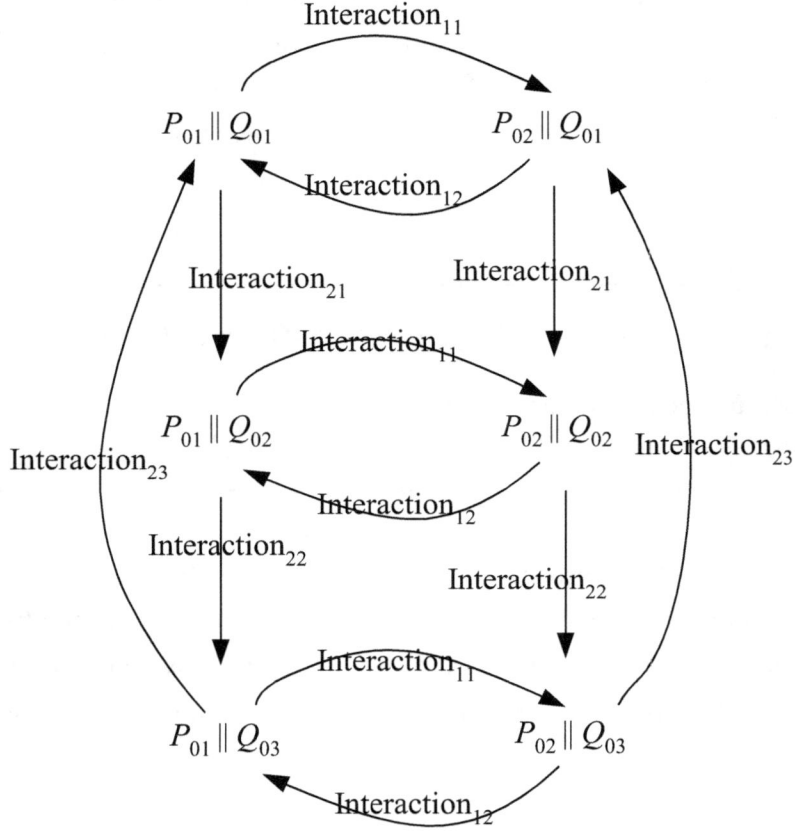

In the transition graph of the *Sweepstakes Sales Promotion* system's multi-queue SBC process, processes $P_{01}\|Q_{01}$, $P_{02}\|Q_{01}$, $P_{01}\|Q_{02}$, $P_{02}\|Q_{02}$, $P_{01}\|Q_{03}$, and $P_{02}\|Q_{03}$ are defined as:

$$P_{01}\|Q_{01} \overset{\text{def}}{=} \text{Interaction}_{11} \bullet (P_{02}\|Q_{01}) + \text{Interaction}_{21} \bullet (P_{01}\|Q_{02})$$

$$P_{02}\|Q_{01} \overset{\text{def}}{=} \text{Interaction}_{12} \bullet (P_{01}\|Q_{01}) + \text{Interaction}_{21} \bullet (P_{02}\|Q_{02})$$

$$P_{01}\|Q_{02} \overset{\text{def}}{=} \text{Interaction}_{11} \bullet (P_{02}\|Q_{02}) + \text{Interaction}_{22} \bullet (P_{01}\|Q_{03})$$

$$P_{02}\|Q_{02} \overset{\text{def}}{=} \text{Interaction}_{12} \bullet (P_{01}\|Q_{02}) + \text{Interaction}_{22} \bullet (P_{02}\|Q_{03})$$

$$P_{01}\| Q_{03} \overset{\text{def}}{=} \text{Interaction}_{11} \bullet (P_{02}\|Q_{03}) + \text{Interaction}_{23} \bullet (P_{01}\|Q_{01})$$

$$P_{02}\|Q_{03} \overset{\text{def}}{=} \text{Interaction}_{12} \bullet (P_{01}\|Q_{03}) + \text{Interaction}_{23} \bullet (P_{02}\|Q_{01})$$

Multi-Queue SBC Process Algebra Defines the Multi-Queue Interactions Scheduling

Examining both the multi-queue model for interactions scheduling and multi-queue SBC process algebra, we find out that both of them generate the same interactions execution sequence.

Therefore, we conclude that multi-queue SBC process algebra defines the multi-queue interactions scheduling.

PART XI: INFINITE-QUEUE MODEL FOR INTERACTIONS SCHEDULING

Approach of Infinite-Queue Model

Infinite-queue model (IQM) adopts an infinite queues interactions scheduling algorithm. The number of queues is infinite. All queues are treated equally, without any preference.

Whenever the yth execution of the xth interaction flow diagram, IFD_x, is ready, a completely new queue, $IFDEXECUTION_{xy}\_QUEUE$, will be prepared and all interactions of this yth execution such as $interaction_{xy1}$, $interaction_{xy2}$,..., $interaction_{xyN}$ will be one-by-one added at the end of the $IFDEXECUTION_{xy}\_QUEUE$ queue. Since there are infinite executions of an interaction flow diagram, there will be an infinite number of new queues.

The array *Ready_Head* has one entry for each queue, with that entry pointing to the interaction at the head of the queue.

124

Ready_Head

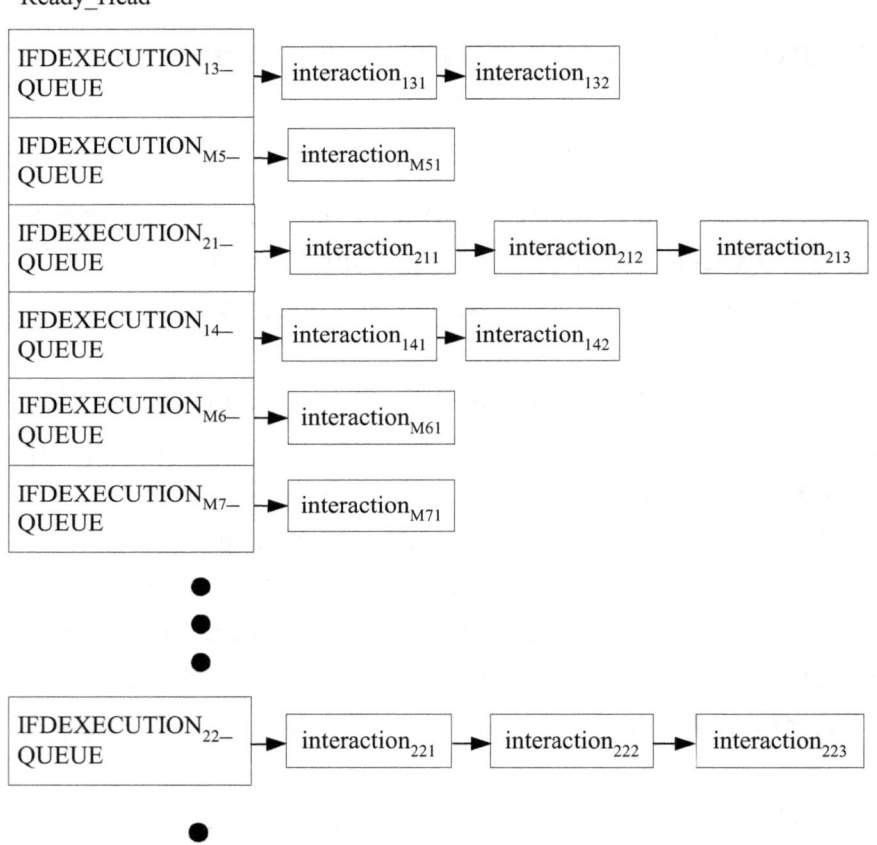

Given the queue structures just described, the IQM scheduling algorithm is simple. It just randomly chooses a queue that is not empty and picks the interaction at the head of that queue. If all the queues are empty, then the idle routine will be executed.

Features of Infinite-Queue Model

The sequence generated by the infinite-queue model for interactions scheduling algorithm possesses the following characteristics:

(1) It is a fair scheduling. Every executable interaction will and shall eventually be executed.

(2) If $e < f$, then the $interaction_{xye}$ must be executed before the $interaction_{xyf}$.

PART XII: INFINITE-QUEUE SBC
PROCESS ALGEBRA

Language Constructs of Infinite-Queue SBC Process Algebra

The set of infinite-queue SBC process for systems architecture is defined by the following BNF grammar:

(1) <Infinite-Queue_SBC_Process_of_Systems_Architecture> ::=
     <Parallel_of_!IFD>

(2) <Parallel_of_!IFD> ::=   *STOP*
   | <!IFD> "‖" <Parallel_of_!IFD>

(3) <!IFD> ::=   <IFD>
   | <IFD> "‖" <!IFD>

(4) <IFD> ::=
   <Type_1_Expression> "●" <Zero_Or_More_Expressions>

(5) <Zero_Or_More_Expressions> ::=   *STOP*
   | <Type_1_Or_2_Expression> "●" <Zero_Or_More_Expressions>

(6) <Type_1_Or_2_Expression>   ::= <Type_1_Expression>
   | <Type_2_Expression>

(7) <Type_1_Expression>   ::= <Type_1_Interaction>
   | <Condition> <Type_1_Interaction>
    {"+" <Condition> <Type_1_Interaction>}

(8) <Type_2_Expression>   ::= <Type_2_Interaction>
   | <Condition> <Type_2_Interaction>
    {"+" <Condition> <Type_2_Interaction>}

(9) <Type_1_Interaction> ::= <Actor> <Operation_Call_Or_Return>
    <Operation_Call_Or_Return_Formula> <Component>

(10) <Type_2_Interaction> ::= <Component> <Operation_Call_Or_Return>
    <Operation_Call_Or_Return_Formula> <Component>

Rule 1 describes that the parallel of all replicated interaction flow diagrams (i.e. Parallel_of_!IFD), in which each interaction flow diagram may replicate itself a countably infinite times (i.e. !IFD), defines the infinite-queue SBC process for systems architecture.

Rule 1
<Infinite-Queue_SBC_Process_for_Systems_Architecture> ::=   <Parallel_of_!IFD>

Rule 2 describes that we use either a) a null process (i.e. *STOP*), or b) the parallel composition (i.e. ‖) composing all replicated interaction flow diagrams (i.e. !IFD), to define the parallel of all replicated interaction flow diagrams (i.e. Parallel_of_!IFD).

Rule 2
<Parallel_of_!IFD> ::=   *STOP* | <!IFD> "‖" <Parallel_of_!IFD>

Rule 3 describes that the parallel composition (i.e. ∥) of a countably infinite number of an interaction flow diagram (i.e. IFD) defines the replicated interaction flow diagram (i.e. !IFD).

```
Rule 3

<!IFD>  ::=
            <IFD>
        |   <IFD>  "∥"  <!IFD>
```

Rule 4 describes that an interaction flow diagram (i.e. IFD) consists of a type_1 expression (i.e. Type_1_Expression), followed by a sequential composition (i.e. ●), and followed by zero or more expressions (i.e. Zero_Or_More_Expressions).

```
Rule 4

<IFD>  ::=
 <Type_1_Expression> "●" <Zero_Or_More_Expressions>
```

Rule 5 describes that zero or more expressions (i.e. Zero_Or_More_Expressions) either a) are a null process (i.e. *STOP)*, or b) consist of a type_1_or_2 expression (i.e. Type_1_Or_2_Expression), followed by a sequential composition (i.e. ●), and followed by zero or more expressions (i.e. Zero_Or_More_Expressions).

Rule 5

<Zero_Or_More_Expressions> ::=
 	
    S*TOP*
 	
    | <Type_1_Or_2_Expression> "●" <Zero_Or_More_Expressions>

Rule 6 describes that the type_1_or_2 expression (i.e. Type_1_Or_2_Expression) is either a type_1 expression (i.e. Type_1_Expression) or a type_2 expression (i.e. Type_2_Expression).

Rule 6
<Type_1_Or_2_Expression> ::=   <Type_1_Expression> | <Type_2_Expression>

Rule 7 describes that the type_1 expression (i.e. Type_1_Expression) is either a type_1 interaction (i.e. Type_1_Interaction) or a conditional type_1 interaction (i.e. one-or-more-armed conditional expression of Type_1_Interaction).

Rule 7

<Type_1_Expression>   ::= <Type_1_Interaction>
 | <Condition> <Type_1_Interaction>
   {"+" <Condition> <Type_1_Interaction>}

Rule 8 describes that the type_2 expression (i.e. Type_2_Expression) is either a type_2 interaction (i.e. Type_2_Interaction) or a conditional type_2 interaction (i.e. one-or-more-armed conditional expression of Type_2_Interaction).

Rule 8

<Type_2_Expression>   ::= <Type_2_Interaction>
 | <Condition> <Type_2_Interaction>
   {"+" <Condition> <Type_2_Interaction>}

Rule 9 describes that an actor interacting with a component defines the type_1 interaction.

Rule 9
<Type_1_Interaction> ::= <Actor> <Operation_Call_Or_Return> <Operation_Call_Or_Return_Formula> <Component>

Rule 10 describes that a component interacting with another component defines the type_2 interaction.

Rule 10
<Type_2_Interaction> ::= <Component> <Operation_Call_Or_Return> <Operation_Call_Or_Return_Formula> <Component>

Transitional Semantics of Infinite-Queue SBC Process Algebra

We assume an infinite set Δ of type_1_or_2 interactions, and use a, b to range over Δ. Further, we let Φ be the set of process Constants, and use A, B to range over Φ. We let Π be the set of processes, and use P, Q to range over Π. We let Ψ be the set of process expressions, and use E, F to range over Ψ.

Entity set	Entity name	Type of entity
Δ	a, b,...	type_1_or_2 interactions
Φ	A, B,...	process Constants
Π	P, Q,...	processes
Ψ	E, F,...	process expressions

In giving meaning to the infinite-queue SBC process algebra, we shall use the following labelled transition system (LTS)

$$(\Psi, \Delta, \rightarrow)$$

which consists of a set Ψ of process expressions, a set Δ of "type 1 or 2 interactions", and a transition relation $\rightarrow \subseteq \Psi X \Delta X \Psi$ where $(E_i, a, E_j) \in \rightarrow$ is denoted by $E_i \xrightarrow{a} E_j$.

The semantics for Ψ consists in the transition rules of each transition relation $\rightarrow$ over $\Psi X \Delta X \Psi$. These transition rules will follow the structure of expressions.

We give the complete set of transition rules; the names Prefix, Parallel, and Constant indicate that the rules are associated respectively with Prefix, Parallel Composition and with Constants.

Prefix	$$\frac{\quad}{a\bullet E \xrightarrow{a} E}$$
Parallel$_1$	$$\frac{E \xrightarrow{a} E'}{E \parallel F \xrightarrow{a} E' \parallel F}$$
Parallel$_2$	$$\frac{F \xrightarrow{a} F'}{E \parallel F \xrightarrow{a} E \parallel F'}$$
Constant	$$\frac{P \xrightarrow{a} P'}{A \xrightarrow{a} P'} \quad (A \overset{def}{=} P)$$

The rule for Prefix can be read as follows: Under any circumstances, we always infer $a \bullet E \xrightarrow{a} E$. That is, an expression, with an interaction prefixed to it, will use this interaction to accomplish the transition.

$$\frac{\rule{3cm}{0.4pt}}{a \bullet E \xrightarrow{a} E}$$

There are two transition rules for parallel composition. Rule Parallel$_1$ indicates that from $E \xrightarrow{a} E'$ we shall infer $E \| F \xrightarrow{a} E' \| F$.

$$\frac{E \xrightarrow{a} E'}{E \| F \xrightarrow{a} E' \| F}$$

Rule Parallel$_2$ indicates that from $F \xrightarrow{a} F$" we shall infer

$E\|F \xrightarrow{a} E\|F$".

$$\frac{F \xrightarrow{a} F'}{E \| F \xrightarrow{a} E \| F'}$$

The rule for Constants can be read as follows: the rule of Constants asserts that each Constant has the same transitions as its defining expression.

$$\frac{P \xrightarrow{a} P'}{A \xrightarrow{a} P'} \quad (A \overset{\text{def}}{=\!=} P)$$

PART XIII: INFINITE-QUEUE INTERACTIONS SCHEDULING VS. INFINITE-QUEUE SBC PROCESS ALGEBRA

Infinite-Queue Interactions Scheduling for the Sweepstakes Sales Promotion System

Sweepstakes Sales Promotion system includes two interaction flow diagrams: IFD_1 and IFD_2.

Whenever the yth execution of the xth (x = 1 to 2) interaction flow diagram, $IFD_{x \ (x = 1 \ to \ 2)}$, of the *Sweepstakes Sales Promotion* system is ready, a completely new queue, $IFDEXECUTION_{xy}\_QUEUE$, will be prepared and all interactions of this yth execution such as $interaction_{xy1}$, $interaction_{xy2},...,$ $interaction_{xyN}$ will be one-by-one added at the end of the $IFDEXECUTION_{xy}\_QUEUE$ queue. Since there are infinite executions of an interaction flow diagram, there will be an infinite number of new queues.

The array *Ready_Head* has one entry for each queue, with that entry pointing to the interaction at the head of the queue.

144

Ready_Head

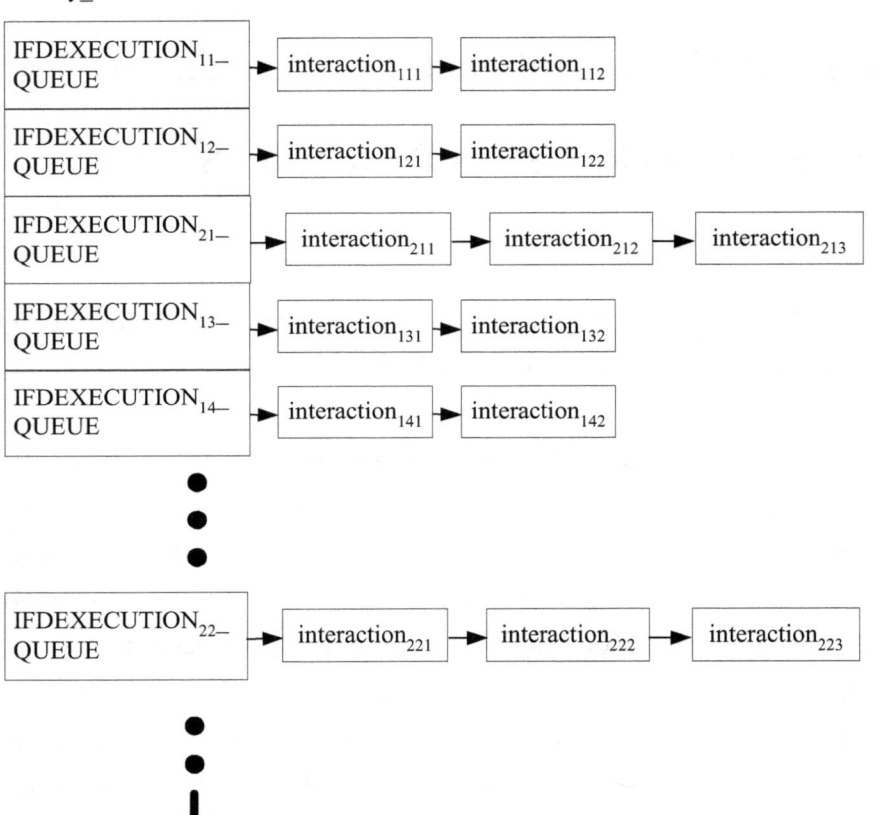

Infinite-Queue SBC Process of the Sweepstakes Sales Promotion System

We draw the infinite-queue SBC process algebra Backus-Naur Form tree of the *Sweepstakes Sales Promotion* system as follows:

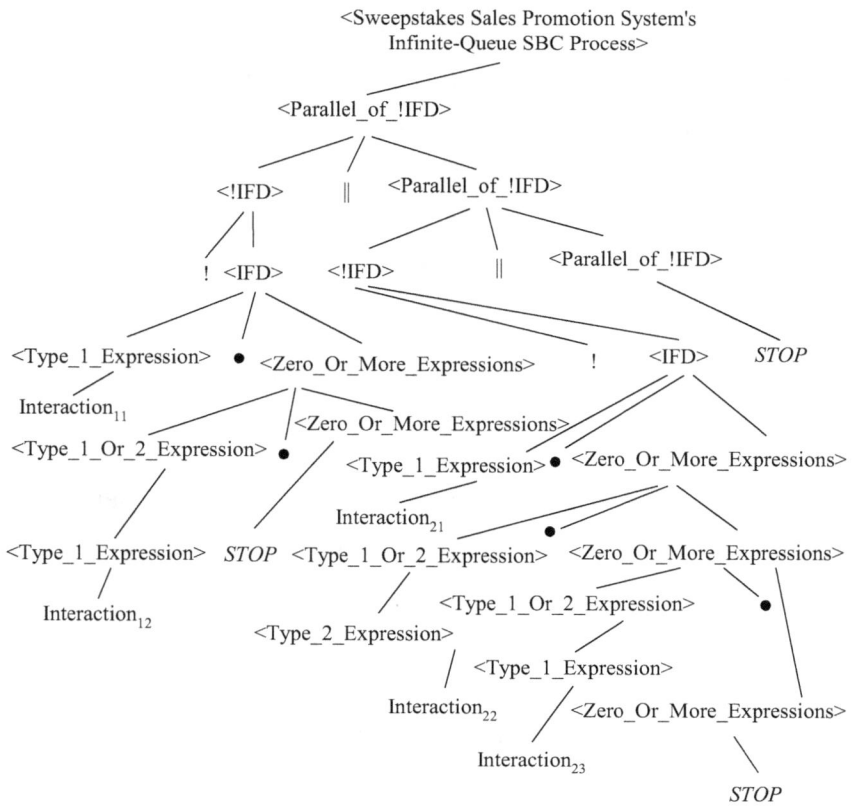

IFD$_1$ describes the 1st interaction flow diagram, i.e. *Get_Sweepstake_Number* behavior, of the *Sweepstakes Sales Promotion* system. IFD$_1$ is syntactically represented as "Interaction$_{11}$●Interaction$_{12}$●*STOP*".

IFD$_1$ =

Interaction$_{11}$● Interaction$_{12}$●*STOP*

IFD$_2$ describes the 2nd interaction flow diagram, i.e. *Draw_Out_the_Winners_List* behavior, of the *Sweepstakes Sales Promotion* system. IFD$_2$ is syntactically represented as "Interaction$_{21}$●Interaction$_{22}$●Interaction$_{23}$●*STOP*".

IFD$_2$ =

Interaction$_{21}$● Interaction$_{22}$● Interaction$_{23}$●*STOP*

We define process P_{01} as the IFD$_1$ process. Thereafter, semantics of the IFD$_1$ process is demonstrated by the following transition graph.

$$P_{01} \xrightarrow{\text{Interaction}_{11}} P_{02} \xrightarrow{\text{Interaction}_{12}} P_{03}$$

In the transition graph of the IFD_1 process, processes P_{01}, P_{02}, and P_{03} are defined as:

$$P_{01} \stackrel{\text{def}}{=} \text{Interaction}_{11} \bullet P_{02}$$

$$P_{02} \stackrel{\text{def}}{=} \text{Interaction}_{12} \bullet P_{03}$$

$$P_{03} \stackrel{\text{def}}{=} STOP$$

We define process Q_{01} as the IFD_2 process. Thereafter, semantics of the IFD_2 process is demonstrated by the following transition graph.

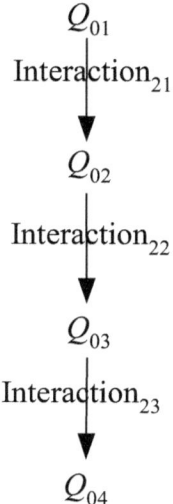

In the transition graph of the IFD$_2$ process, processes Q_{01}, Q_{02}, Q_{03}, and Q_{04} are defined as:

$$Q_{01} \overset{def}{=\!=} \text{Interaction}_{21} \bullet Q_{02}$$

$$Q_{02} \overset{def}{=\!=} \text{Interaction}_{22} \bullet Q_{03}$$

$$Q_{03} \overset{def}{=\!=} \text{Interaction}_{23} \bullet Q_{04}$$

$$Q_{04} \overset{def}{=\!=} STOP$$

We define process $P_{01} \| Q_{01}$ as the IFD$_1 \|$IFD$_2$ process. Thereafter, semantics of the IFD$_1 \|$IFD$_2$ process is demonstrated by the following transition graph.

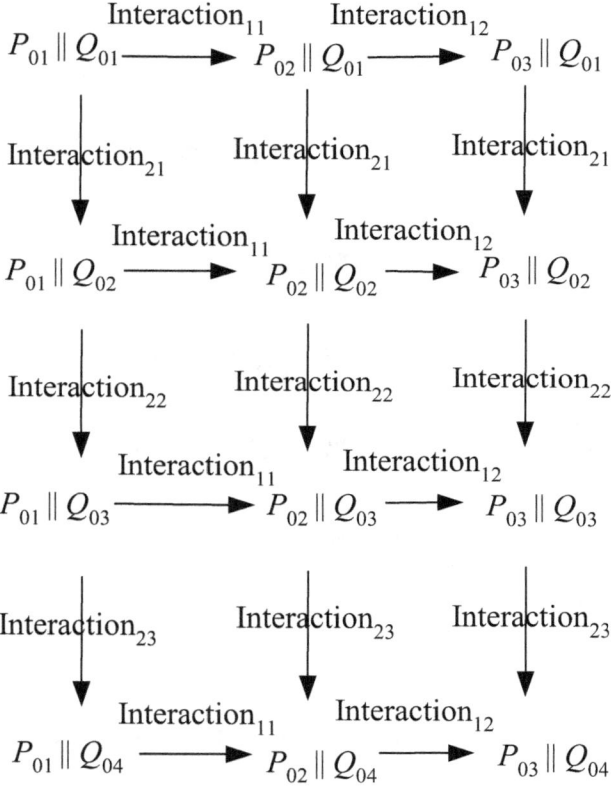

In the transition graph of the $IFD_1 \| IFD_2$ process, processes $P_{01}\|Q_{01}$, $P_{02}\|Q_{01}$, $P_{03}\|Q_{01}$, $P_{01}\|Q_{02}$, $P_{02}\|Q_{02}$, $P_{03}\|Q_{02}$, $P_{01}\|Q_{03}$, $P_{02}\|Q_{03}$, $P_{03}\|Q_{03}$, $P_{01}\|Q_{04}$, $P_{02}\|Q_{04}$, and $P_{03}\|Q_{04}$ are defined as:

$$P_{01}\|Q_{01} \overset{\text{def}}{=\!=} \text{Interaction}_{11} \bullet (P_{02}\|Q_{01}) + \text{Interaction}_{21} \bullet (P_{01}\|Q_{02})$$

$$P_{02}\|Q_{01} \overset{\text{def}}{=\!=} \text{Interaction}_{12} \bullet (P_{03}\|Q_{01}) + \text{Interaction}_{21} \bullet (P_{02}\|Q_{02})$$

$$P_{03}\|Q_{01} \overset{\text{def}}{=\!=} \text{Interaction}_{21} \bullet (P_{03}\|Q_{02})$$

$$P_{01}\|Q_{02} \overset{\text{def}}{=\!=} \text{Interaction}_{11} \bullet (P_{02}\|Q_{02}) + \text{Interaction}_{22} \bullet (P_{01}\|Q_{03})$$

$$P_{02}\|Q_{02} \overset{\text{def}}{=\!=} \text{Interaction}_{12} \bullet (P_{03}\|Q_{02}) + \text{Interaction}_{22} \bullet (P_{02}\|Q_{03})$$

$$P_{03}\|Q_{02} \overset{\text{def}}{=\!=} \text{Interaction}_{22} \bullet (P_{03}\|Q_{03})$$

$$P_{01}\| Q_{03} \overset{\text{def}}{=\!=} \text{Interaction}_{11} \bullet (P_{02}\|Q_{03}) + \text{Interaction}_{23} \bullet (P_{01}\|Q_{04})$$

$$P_{02}\|Q_{03} \overset{\text{def}}{=\!=} \text{Interaction}_{12} \bullet (P_{03}\|Q_{03}) + \text{Interaction}_{23} \bullet (P_{02}\|Q_{04})$$

$$P_{03}\|Q_{03} \overset{\text{def}}{=\!=} \text{Interaction}_{23} \bullet (P_{03}\|Q_{04})$$

$$P_{01}\| Q_{04} \overset{\text{def}}{=\!=} \text{Interaction}_{11} \bullet (P_{02}\|Q_{04})$$

$$P_{02}\|Q_{04} \overset{\text{def}}{=\!=} \text{Interaction}_{12} \bullet (P_{03}\|Q_{04})$$

$$P_{03}\|Q_{04} \overset{\text{def}}{=\!=} STOP$$

The *Sweepstakes Sales Promotion* system's infinite-queue SBC process is syntactically represented as $(!IFD_1)\|(!IFD_2)$ which equals to $!(IFD_1\|IFD_2)$ which also equals to "(Interaction$_{111}$•Interaction$_{112}$•*STOP*) $\|$
(Interaction$_{211}$•Interaction$_{212}$•Interaction$_{213}$•*STOP*) $\|$
(Interaction$_{121}$•Interaction$_{122}$•*STOP*) $\|$
(Interaction$_{221}$•Interaction$_{222}$•Interaction$_{223}$•*STOP*) $\|$
(Interaction$_{131}$•Interaction$_{132}$•*STOP*) $\|$
(Interaction$_{231}$•Interaction$_{232}$•Interaction$_{233}$•*STOP*) .... $\|$
(Interaction$_{1\infty1}$•Interaction$_{1\infty2}$•*STOP*) .... $\|$
(Interaction$_{2\infty1}$•Interaction$_{2\infty2}$•Interaction$_{2\infty3}$•*STOP*)".

Sweepstakes Sales Promotion system's Infinite-Queue SBC Process $\overset{def}{=\!=}$

(Interaction$_{111}$•Interaction$_{112}$•*STOP*)
  $\|$
(Interaction$_{211}$•Interaction$_{212}$•Interaction$_{213}$•*STOP*)
  $\|$
(Interaction$_{121}$•Interaction$_{122}$•*STOP*)
  $\|$
(Interaction$_{221}$•Interaction$_{222}$•Interaction$_{223}$•*STOP*)
  $\|$
(Interaction$_{131}$•Interaction$_{132}$•*STOP*)
  $\|$
(Interaction$_{231}$•Interaction$_{232}$•Interaction$_{233}$•*STOP*) ....
  $\|$
(Interaction$_{1\infty1}$•Interaction$_{1\infty2}$•*STOP*) ....
  $\|$
(Interaction$_{2\infty1}$•Interaction$_{2\infty2}$•Interaction$_{2\infty3}$•*STOP*)

Infinite-Queue SBC Process Algebra Defines the Infinite-Queue Interactions Scheduling

Examining both the infinite-queue model for interactions scheduling and infinite-queue SBC process algebra, we find out that both of them generate the same interactions execution sequence.

Therefore, we conclude that infinite-queue SBC process algebra defines the infinite-queue interactions scheduling.

www.ingramcontent.com/pod-product-compliance
Lightning Source LLC
Chambersburg PA
CBHW051918170526
45168CB00001B/445